EZEKIEL'S CHARIOT OF

CW00567671

Also from Tiger of the Stripe

The Architectural History of Canterbury Cathedral
The Rev. R. Willis MA, FRS

The Ecclesiastical History of the English People
The Venerable Bede, edited by J. A. Giles and Gerrish Gray

THE SCIENCE OF EZEKIEL'S CHARIOT OF YHWH VISION AS A SYNTHESIS OF REASON AND SPIRIT

Patrick Pullicino

TIGER OF THE STRIPE

2023

This book first published in 2023 by
TIGER OF THE STRIPE
50 Albert Road
Richmond
Surrey TW10 6DP
United Kingdom

tigerofthestripe.co.uk

© 2022, 2023 Patrick Pullicino
All rights reserved

*The text is based on a thesis
successfully submitted to the
Institute of Education,
St Mary's, University of London,
for the award of
Doctor of Philosophy*

ISBN 978-1-904799-74-0

Typeset in the United Kingdom by
Tiger of the Stripe

Contents

למרים הנהר הנצחי שמביא חיים ובריאות

Abstract

Introduction

The Chariot Vision (CV) of Ezekiel 1 has long been a source of fascination for exegetes. Patristic interpreters saw the individual character of the gospels reflected in the four faces of the Living Creatures and the CV as an important link between Old and New Testaments. While the focus of exegesis has largely explored the CV's theological symbolism, its scientific significance has also attracted scholars. Moses Maimonides (1135-1204) held that Ezekiel's CV instances what might be termed 'divine science' as it exemplifies how rationality and revelation co-inhere. Launderville proposes a facilitation of reason by the spirit among the Israelites that connects them with both reality and the cosmic order. Using Ezekiel's ancient context, I explore these premises and how the CV synthesizes reason and spirit and exemplifies links between rationality and revelation.

Method

The thesis reviews readings of the CV within its immediate biblical context, against its background of Mesopotamian understandings of science and cosmology while critically engaging with insights of Christian patristic interpreters, notably Origen and Jerome. The style and structure of the book of Ezekiel, its relationship to the prophet and its use of rational structures in metaphor and symbol are considered. The symbolism of the CV's Living Creatures is explored in the context of Mesopotamian astrology/astronomy and through the key link forged throughout the text between measurement and rationality as a feature of ancient mathematics and pre-science. The three-level hierarchy of the CV is investigated as an exemplar both of Maimonides' rational revelation and Launderville's interaction between spirit and reason.

Findings

The three levels of the CV illuminate our understanding of the relationship between rationality, inspiration and revelation and may therefore rightly be termed 'divine science'. An interaction between spirit and reason are detectable in different 'spheres' of the CV. In this sense, it allows the thesis to suggest a number of hermeneutical possibilities which merit scholarly consideration: (a) that the faces of the Living Creatures symbolize rational and emotional dimensions of decision making; (b) that 'each moved straight ahead' is an analogy for right reason; (c) that 'the spirit of the living creatures was in the wheels' offers insight into the hierarchy of reason; (d) that the vision portends the 'new heart' and 'new spirit' for the Israelites; (e) that the constituent elements of the Chariot Vision imply interconnectivity between cosmological, psychological, epistemological and metrological/scientific experimental domains.

Conclusion

The CV is both spiritually and scientifically profound. Understanding Ezekiel's CV as a synthesis of Reason and Spirit is not only a hermeneutically sound interpretation of its symbolism, but it also has the potential to contribute to contemporary questions regarding the unity of knowledge.

Abbreviations

Ann Indian Acad Neurol *Annals of the Indian Academy of Neurology*

Arq Neuro-Psiquiatr *Arquivos de Neuro-Psiquiatria*

ANET *Ancient Near Eastern Texts*

Astro-ph *Astrophysics*

Behav Brain Sci *Behavioural and Brain Sciences*

BZAW *Beihefte zur Zeitschrift für die alttestamentliche Wissenschaft*

CBR *Community Bible Reading Journal*

Clin Child Fam Psychol Rev *Clinical Child and Family Psychology Review*

CT *Cuneiform Text*

Front Psychol *Frontiers in Psychology*

Front Hum Neurosci *Frontiers in Human Neuroscience*

Hum Neurobiol *Human Neurobiology*

IEEE *Institute of Electrical and Electronics Engineers*

Int J Appl Math Comput Sci *International Journal of Applied Mathematics and Computer Science*

Int J Ment Health Syst *International Journal of Mental Health Systems*

IRE *Institute of Radio Engineers*

JAOS *Journal of the American Oriental Society*

JAMA *Journal of the American Medical Association*

JBL *Journal of Biblical Literature*

JBR *The Journal of the Bible and its Reception*

J Br Astron Assoc *Journal of the British Astronomical Association*

JETS *Journal of Evangelical Theological Society*

J Immigr Minor Health *Journal of Immigrant and Minority Health*

NRSV *New Revised Standard Version*

NTS *New Testament Studies*

PLOS *Public Library of Science*

Psychol Res *Psychological Research*

Psychol Rev *Psychological Review*

Soc Sci Med *Social Science and Medicine*

Sci Rep *Scientific Reports*

ST *Summa Theologiae*

TMSJ *The Master's Seminary Journal*

Trans Syst Man Cybern *Transactions on Systems, Man, and Cybernetics*

TynBul *Tyndale Bulletin*
VT *Vetus Testamentum*
ZAW *Zeitschrift für die alttestamentliche Wissenschaft*
𝔊 Septuagint
𝔐 Masoretic
* Footnote that clarifies or amplifies the text

Introduction: Divine Science in Ezekiel's Chariot Vision

The initial manifestation and description of Ezekiel's *Merkavah* or Chariot Vision is unique in Scripture. It is a vision which is at once complex, strange and almost dreamlike in parts, but compelling.[1] It is compelling partly because it presents to us the 'Glory of God' in theophanic and mysterious detail and partly because of its ordered structure. It has long attracted the attention of exegetes. The Chariot Vision was a favoured focus of study, particularly the Living Creatures and wheels. Origen sees Christ and the Church foreshadowed in Ezekiel and this is true also of his reading of the Chariot Vision.[2] St Jerome, in his lengthy exposition on the Chariot Vision, treats the vision as revelation that is obscure and needs interpretation.[3] To Maimonides, the Chariot Vision is revelatory and he called it *al-'ilm al-ilāhi* (العلم الالهي) which is translated as 'divine science,'[4] or a presentation of the divine in a rational way.

---

[1] Blenkinsopp, J. (1990), 'Ezekiel,' *Interpretation: A Journal of Bible and Theology*, Louisville: John Knox Press, p.18.

[2] Origen (2010), *Homilies 1-14 on Ezekiel*, trans Sheck, T. P. , New York: Newman Press, p. 45.

[3] St Jerome (2017), *Commentary on Ezekiel, Ancient Christian Writers*, New York: The Newman Press, p. 34.

[4] Davies, D. (2011), *Method and Metaphysics in Maimonides Guide for the Perplexed*. Oxford: Oxford University Press, p. 1; 'In the Maimonidean tradition... the scope of the divine science focuses on the entire order of existence in order to provide a comprehensive picture of the world, the understanding of which, in the eyes of the philosophers, leads

The spirit is a unifying theme across the Book of Ezekiel.[5] Maimonides sees the action of the spirit at a cosmic level, with the Chariot Vision as a parallel of *ma'aseh berešit* (מעשה בראשית) or Genesis 1, with the Chariot Vision's crystal firmament and tiered structure giving a rational representation of cosmic spheres.[6] For Launderville the spirit is the reality that connects humans with the cosmic order, by correcting the function of the Israelites' wayward hearts.[7] Jeremiah had introduced the importance of the heart for the new covenant,[8] and Ezekiel confirms this. Jerome recognised the heart as the traditional loc-

---

to human perfection.' Kreisel, H. (2015), *Judaism as Philosophy: Studies in Maimonides and the Medieval Jewish Philosophers of Provence,* Boston: Academic Studies Press, p. 260; Aquinas says that Divine Science is a term for the 'part of philosophy that is called theology'. It contains conclusions reached by rational insights. He states that Divine Science is the only and simple, law of all things: *Summa Theologiae* I, I, 3, *ad* 2; In contrast, Aquinas states that *Sacra Doctrina* contains the implications and origins of revealed truths. Although it is a science (*ST* I, I, 2) it only includes revealed doctrine and not theological conclusions reached by rational or philosophical insights. *Sacra Doctrina* is part of Sacred Science, which is the study of revealed truth. Davies B. (1990), 'Is Sacra Doctrina theology?,' *New Blackfriars,* **71** (836), pp. 141–47.

[5] Launderville D. (2004), 'Ezekiel's Throne-Chariot Vision: Spiritualizing the Model of Divine Royal Rule,' *Catholic Biblical Quarterly* **66** (3), pp. 361–377; p. 375; Merriam Webster's Dictionary gives 14 different uses of 'spirit': *https://www.merriam-webster.com/dictionary/spirit.* The meaning of 'spirit' as used in this thesis are those encompassed by the Hebrew word *rúaḥ* in the Bible and particularly in Ezekiel as discussed in chapter 8.

[6] 'the Spirit of God was moving over the face of the waters, (Ge 1:2); 'let there be a firmament in the midst of the waters', (Ge 1:6); Maimonides sees the Chariot Vision as a parable of the description of the Cosmos in Genesis; Davies, D. (2011), *Ibid.,* pp. III, 135; Maimonides sees the Accounts of Creation and of the Chariot as both sciences dealing with the same entities: separate intellects, celestial spheres and earthly elements. He sees the Aristotelian view of the structure of the world as the key to understanding the Chariot vision with the Living Creatures representing celestial spheres: Kreisel, H. (2015), *Ibid.,* pp. 244, 248.

[7] Launderville, D.F. (2007) *Spirit and Reason,* Waco: Baylor University Press, p. 93.

[8] Je 31:33 'I will put My law within them and write it on their heart'

ation of thought and feelings,[9] and in his seminal work on Ezekiel, Launderville has convincingly advanced the 'heart' as an Ezekelian metaphor for a crucial synthesis of spirit and reason.

Revelation is an overt motivation in Ezekiel: a stressed motif in the Restoration Oracle (Ez 36:16–38) is that God's actions in human history are driven by revelatory aims – that his people and the world may know he is YHWH.[10] Launderville demonstrates that this revelation is spirit-driven and is 'embodied', as it enables the workings of the human heart. The heart is where reason is guided by the spirit in order that the Israelites become living symbols of YHWH.[11] Since the Chariot Vision is the 'keel' of Ezekiel it would therefore be entirely consistent for a crucial connection between reason and spirit to be deeply embedded within it and moreover that the synthesis of reason and spirit, the 'heart', should be deliberately represented. This thesis will address both issues within the Chariot Vision. In doing so it will bring into perspective the cosmological, psychological, epistemological and metrological themes which are integral to understanding the significance of the text.

In the close reading of the Chariot Vision which follows, this thesis proposes to discern the presence of an interaction between reason and spirit throughout the Chariot Vision. The Chariot Vision itself is hierarchical and its structured tiers evoke rationality:[*12] rationality that interacts with the spirit that drives the

---

[9]   St Jerome (2017), *Ibid.*, p. 33; In Hebrew, there is no separate location for noetic activity and the heart is the place in which noetic and affective activities converge: Launderville, D. F. (2007), *Ibid.* p.54.

[10]   Block, D. I. (1998), *The Book of Ezekiel Chapters 25–48*, Grand Rapids: Eerdmans, p. 366.

[11]   Launderville, D. F. (2007), *Spirit and Reason*, Waco: Baylor University Press, p. 348.

[12]   Cook, S. L. (2004) 'Cosmos, *Kabod* and Cherub: Ontological and Epistemological Hierarchy in Ezekiel,' in *Ezekiel's Hierarchical World: Wrestling with a Tiered Reality*, eds C. L. Patton and S. L. Cook, Atlanta: Society of Biblical Literature, 179–198, p. 185; Reason is the justification for an action or event. Reason is also the power of the mind to think, understand, and form judgements logically. Rationality is the quality of being based on or in accordance with reason or logic, i.e. with valid reason. Rationality originates from reason, but reason does not originate from rationality,

Chariot 'for the spirit of the living creatures was in the wheels' (Ez 1:20; 10:17). The action of the spirit in the new heart promised by YHWH, is the restoration of a harmonious relationship between thinking and feeling,[13] that will cause the Israelites to 'be careful to observe my ordinances' (Ez 36:26–27). This echoes a rectitude in the Living Creatures that, when guided by the spirit, 'each moved straight ahead' and suggests that reason and spirit are crucial cornerstones of the Chariot Vision. In all three tiers of the Chariot Vision, understanding the vision as a synthesis/synergy of Reason and Spirit, will assist us to appreciate the layers of its original meaning and how these insights have contemporary relevance.

The study commences therefore with two essential *contextual chapters*: the thesis will first look at the prophet Ezekiel and the structure of the Book of Ezekiel and then review the relevant literature on the Chariot Vision. Following this, the *three tiers of the Chariot Vision* will be looked at in detail, particularly identifying connections with the spirit. This will include a review of the Mesopotamian and Egyptian *Sitz im Leben* of the faces of the Living Creatures and the Hebrew etymology of the two words for wheel used in Ezekiel: *ôpan* (אוֹפָן) and *galgal* (גַּלְגַּל).

The analysis continues as rationality is explored further through the ubiquitous phenomenon of measurement in Ezekiel, and its counterpart in the Chariot Vision (Chapter 6). Implicit hierarchies, including the epistemology of the Chariot Vision and the links between rationality, science and revelation are considered (Chapters 9 and 12). In Chapter 8, the thesis focuses on *Spirit in Ezekiel and the Chariot* and *The Interaction of Reason and Spirit* is then detailed (Chapter 11). The thesis includes specific consideration of the use of metaphor and symbolism, which is central to the living symbolism of the Israelites, in whom the spirit has correctly balanced reason (Chapter 7). The

---

(*https://www.merriam-webster.com/dictionary/reason /rationality*). Reason will be analysed in further detail in the Chapter 11.

[13]   Launderville, D. F. (2007), *Ibid.*, p. 348.

movements of the chariot, which also enfold a balancing of reason and spirit, are explored in Chapter 10 which brings together elements from the analytical core of the thesis. In particular, Origen's Christo-centric focus of Ezekiel is explored through the movements of the Chariot.

It will be seen that the fecundity of this reading has ramifications for our understanding of Ezekiel in its ancient context but also for our contemporary appreciation of the text. The methodological approach adopted by this thesis breaks new ground in demonstrating how the Chariot Vision has relevance to perennial and fundamental questions regarding God, humanity, revelation and the intelligibility of the cosmos.

Chapter 1

Background to Ezekiel and the Book

Overview of chapter

This chapter will first give a brief introduction to Ezekiel stressing areas to be covered in the thesis. The rationale for using the Masoretic Hebrew text source is then discussed. A note on the 'torah' of Ezekiel, as it reflects the Mosaic Torah follows. The chapter then turns to Ezekiel: his personality, grounds for single authorship of the text versus redaction over time; potential influences on Ezekiel's writing style including an analysis of illnesses attributed to him; his development as a powerful prophet and finally evidence that the book was written by Ezekiel himself.

Introductory Overview to the Book of Ezekiel

Ezekiel (623–570 BC) is chronologically the third of the major prophets. At the age of 30, (593 BC) Ezekiel started as a visionary and prophet among the Exiles in Babylon, after the first siege of Jerusalem (598–597 BC). Ezekiel is a complex book with multiple themes and Ezekiel prophesies particularly about the future destruction of Jerusalem and the Temple and its aftermath. The subject matter of Ezekiel is wider than the other prophets and extends from basic strict personal moral codes to complex and dream-like visions often incorporating movement, that are described in great detail. The mood spans from unremittingly harsh punishment for sins to a gentle message of hope for the repentant and for the future. The Book of Ezekiel contains several detailed and vivid visions, two of which span several chapters: The Chariot Vision and Idealised Temple Vision. The 'Glory of the Lord' or Chariot (*Merkavah*) Vision is seen on five separate occasions by Ezekiel. This complex mystical vision has resisted clear interpretation but has become a defining feature of the book and has been a focal point for Jewish mysticism.[14] Ezekiel's visions have an eschatological flavour, and encompass the timeless glory of God, the future restoration of Jerusalem and its temple as well as the raising to life of the dead, in the Vision of the Dry Bones. In addition to prophesying, Ezekiel was a 'priest without-a-Temple'. He is set as a watchman for Israel and in a priestly manner, he repeatedly castigates Israel for apostasy. Despite this he has a message of consolation and of restor-ation of Jerusalem and that of a new spirit for the people. Symbolically, the Israelites have a heart of stone, and YHWH will give them a heart of flesh (Ez 36:26). YHWH calls sinners to inner conversion, to careful observance of his laws and covenant. Ezekiel gives personal laws of the imperative to continuing righteousness and a mortal warning for those who persist in evil or for righteous people who turn to evil. Two Halakhic laws stem from Ezekiel,[15] (Ez 39:15; 44:9) and Ezekiel is the only book of the Bible in which objective standards of measurement are given (Ez 45:10-15).

One of the main underlying themes of Ezekiel is of God's transcendent glory and perfect holiness and the deep respect that is due to Him. Ezekiel is a major

---

[14] The *Merkavah* vision has for generations been a focus of Jewish meditation, holding the deepest secret of the Torah, became a way of seeking YHWH through prayer and mystical elevation of the mind to God. Davies, D. (2011), *Method and Metaphysics in Maimonides Guide for the Perplexed*, Oxford: Oxford University Press, p. 106.

[15] Sweeney, M.A. (2014), *Reading Prophetic Books: Form, Intertextuality, and Reception in Prophetic and Post Biblical Literature*, Tubingen: Mohr Siebeck, p. 363.

early exponent of eschatological literature[16] (Ez 38-39) and the book contains a message of belief and hope in final salvation. The style and content of writing varies from seemingly chaotic descriptions of visions to extended oracles and legal reflections and to prolonged passages of repetitive instruction and of extreme castigation with use of sexual analogies. The book can be divided into three main stages of the prophet's ministry:[17] Oracles of Judgment (Ez 1-24), Oracles against Foreign Nations (Ez 25-32) and Oracles of Salvation (Ez 33-37). The final section of the book (Ez 38-48) is the prolonged Vision of the Idealised Temple and related topics.

There was much dispute among Jews about putting Ezekiel into the *Tanakh* (the Hebrew canon). The main concerns were: (a) that some of his prophecies did not come about, (particularly the prophecy of the slaughter in Jerusalem, which was not reported by Jeremiah who was alive at the time of the fall of Jerusalem; his prophecy of the fall of Tyre was not fully fulfilled until 250 years later[*18]; (b) His criticism of Israel's apostasy and depravity are extreme and unrelenting and he excludes positive aspects of Jewish history; (c) His detailed sexualized imagery, were said to be unbecoming of a prophet, and his sometimes extreme castigation of women unnecessary (Ez 17). It is said that it was largely because of his visions, especially the *Merkavah* Vision, that he was not excluded from the *Tanakh*.[19]

Note on the Source of the Text Used for this Book

Ezekiel was undoubtedly initially written in Hebrew, but no original texts survive. In the second century BC at the request of Ptolemy II Philadelphus

---

[16] Block, D. I. (2014), *Beyond the River Chebar: Studies in Kingship and Eschatology in the Book of Ezekiel*, Cambridge: James Clarke & Co., p. 95.

[17] Mc Keating, H. (1993), *Ezekiel*, Sheffield: JSOT Press, p. 15.

[18] Zechariah (Ze 9:3-4) and Isaiah (Is 23:15) also prophesied the fall of Tyre.

[19] Rabbi Moshe Reiss Bible Commentator, *Ezekiel: A problematic prophet*, in two parts. *http://moshereiss.org/articles/23_ezekiel.htm* accessed 20/12/2018.

(283-246 BC) about 70 Jewish scholars translated the Hebrew into Greek and the Septuagint (𝕲) became a major standard text of the Bible, particularly for those who could not read Hebrew. Since the last century, Christian exegetes in particular, value this translation as the main source for Ezekiel.[20] The only complete text of Ezekiel is however, the Masoretic text (𝕸), which was written about the sixth century AD and 𝕸 has remained a single standardized text since that time.[21] It is the fruit of transcribers and copiers since the initial pronouncement of the prophet and has inevitable copying errors. However, it is the only extant text in Hebrew. Those who prefer 𝕲 over 𝕸 as a basic Ezekiel text[22] point out that 𝕲 was particularly important in the redaction of 𝕸.[23] Also there have been criticisms of widespread novel emendations in 𝕸.[24] It is clear however, that 𝕲 is a translation which reflected the religious needs of Alexandrian Jews.[25] Hebrew is a very different language from Greek. The phonological system and script of Greek are unlike those of Hebrew, Hebrew is written right to left; Greek nouns are declined and there are very few common words. Hebrew thought is dynamic and most words originate from triradiate verbs. Boman shows that Hebrew thought and language is time-oriented and active and it emphasizes psychology.[26] Greek on the other hand tends to be static and specific. There are many inherent disadvantages therefore in studying a Greek translation of a book like Ezekiel that was originally written in Hebrew.

[20]   Halperin, D.J. (1982), 'Merkabah Midrash in the Septuagint,' *JBL*, pp, 351-363; p. 351.

[21]   Greenberg, M. (1983), *Ezekiel 1-20*, New York: Doubleday, p. 19.

[22]   Zimmerli, W. (1979), *Ezekiel 1*, Philadelphia: Fortress Press, p. 3.

[23]   Cornill, C. H. (1886), *Das Buch des Propheten Ezechiel*, Liepzig: Hinrichs, p. vi.

[24]   Jahn, G. (1905), *Das Buch Ezechiel auf Grund der Spetuaginta hergestellt, übersetzt und kritisch eklärt*, Leipzig: Pfeiffer, p. iii.

[25]   Jellicoe, S. (1968), *The Septuagint and Modern Study*, Oxford: Clarendon, p. 321.

[26]   Boman, T. (1960), *Hebrew Thought Compared with Greek*, Trans. Jules L. Moreau, New York: Norton, p. 23.

Like Greenberg,[27] and Block,[28] this thesis uses the Hebrew text (𝔐) as the text source, but respecting indications of likely additions or omissions compared with 𝔊, which where relevant, are detailed and discussed.

The Torah of Ezekiel

The final section of the Book of Ezekiel has been called the Torah of Ezekiel because of its comparability with the Mosaic Torah.[29] Ezekiel's Torah begins with an opening preamble (Ez 40:1–4) and the text then has three major units: Ez 40:5–43:27 dealing with YHWH coming in to reside in the Temple; Ez 44:1–46:24 which details Israel's response to YHWH's presence in their midst; and Ez 47:1–48:35 containing the apportioning of land to the twelve tribes. These are comparable with: Ex 25:1–40 which begins with the provision for YHWH's residence; the remainder of Exodus and Numbers is the response to YHWH's presence; and Nu 34–35 is the apportioning of land to the twelve tribes. The style and language of Ezekiel are heavily influenced by the Priestly Writings and Holiness Code of the Mosaic Torah,[30] (Le 17–26). The statutes in Ezekiel's Torah (Ez 44–46) however do not incorporate the Holiness Code imperative of social justice (Le 18–26), reinforcing the conclusion that this section is not meant to be an attempt at a reduplication of a comprehensive Torah, but a portrayal of certain central elements contained in it.

There was consensus in the past that chapters 40-48 were so different in style and content from the rest of the book that they represented a post-Ezekiel or postexilic addition.[31] It is now however, generally accepted as consistent with

---

[27]   Greenberg, M. (1983), *Ezekiel 1-20*, New York: Doubleday, p. 19.

[28]   Block, D.I. (1997), *The Book of Ezekiel chapters 1-24*, Grand Rapids: Eerdmans, p. 42.

[29]   Block, D. I. (1998), *The Book of Ezekiel chapters 25-48*, Grand Rapids: Eerdmans, p. 590.

[30]   Block, D. I. (1997), *The Book of Ezekiel chapters 1-24*, Grand Rapids: Eerdmans, p. 40.

[31]   Eichrodt, W. (1970), *Ezekiel*, London: SCM Press, p. 530.

the rest of the book and written by Ezekiel himself.[32] The temple in the vision is clearly idealized.[33] Its exact purpose is still under dispute, but the central theme is the preservation of YHWH's holiness.[34] It has also been suggested that it is an endeavour to map the social configurations of an ideal cultic space.[35] The implications of the prolonged measurement sequences in this section, are discussed below.

Ezekiel's work reflects his personality and experiences

There is an increased understanding of the importance that the background social dimension in Ancient Israel had on prophets.[36] The time and place of Ezekiel's ministry are vital issues in the interpretation of the Book.[37] Among the prophetic books of the Old Testament, the Book of Ezekiel is uniquely linked to the personal experience of Ezekiel for several reasons:

The current consensus is that Ezekiel wrote almost the entire book himself (see below).

1. The son of Buzi, a priest, was well educated and sensitive[38] and a profound theologian.[39] In terms of his work, Greenberg says that the Book of Ezekiel 'is the product of art and intelligent design ...'[40]

32 Greenberg, M. (1984), 'The Design and Themes of Ezekiel's Plan of Restoration,' *Interpretation, A Journal of Bible and Theology*, **38** (2), pp. 181–208.

33 Block, D.I. (1998), *The Book of Ezekiel chapters 25-48*, Grand Rapids: Eerdmans, p. 501.

34 Odell, S. (2005), *Ezekiel*, Smyth & Helwys Bible Commentary, Macon: Smyth & Helwys, p. 482; Block, D. I. (1998), *Ibid.*, p. 582.

35 Smith, J. Z. (1987), *To Take Place: Towards Theory in Ritual*, Chicago: University of Chicago Press, p. 48; Block, D. I. (1998), *Ibid.*, p. 746.

36 Wilson, R.R. (1980), *Prophecy and Society in Ancient Israel*, Minneapolis: Fortress Press, p. 13.

37 Joyce, P. M. (2007), *Ezekiel: A Commentary*, New York: T & T Clark, p. 3.

38 Block, D. I. (1997), *The Book of Ezekiel chapters 1-24*, Grand Rapids: Eerdmans. p. 11.

39 Rad, G. von (2012), *Old Testament Theology, Volume Two: The Theology of Israel's Prophetic Traditions*, Louisville: Westminster John Knox Press, p. 223.

40 Greenberg, M. (1983), *Ezekiel 1-20*, New York: Doubleday, p. 26.

2. Ezekiel was probably training for the priesthood when he was exiled and Ezekiel displays an intense interest in priestly matters.[41] He was exiled to Babylon in the region of Nippur[42] and stayed there until he died, and never had access to the Temple in Jerusalem. Ezekiel was therefore forced to live his priesthood in an unconventional way. With the dramatic appearance of the Chariot Vision in his thirtieth year, Ezekiel must have known that not only did YHWH consider him a priest, but a prophet too[43] and he had to find a way to combine prophecy and priesthood.

3. It has been argued that the Exile had a deep psychological and emotional impact on Ezekiel.[44] The theological dimension of this would be essential for the understanding of Ezekiel.[45]

Authorship and Cohesion of Ezekiel

Debate on the authorship of Ezekiel dates to over 15 centuries ago.[46] Although the book has multiple complex themes it has been accepted as the work of a single author because of its clear underlying structure. As mentioned above, visions, and particularly the Chariot Vision, are an underlying structural link in the book. In addition, the visions and oracles are carefully dated throughout and follow in a chronological order apart from a few inconsistencies.[47] Over the years this structural unity generally convinced commentators that one author, Ezekiel, had written the book.

---

41 Block, D. I. (1997), *Ibid.,* p. 88.

42 Block, D. I. (1997), *Ibid.,* p. 84.

43 Ez 2:5 'and they will know that there has been a prophet among them'

44 Smith-Christopher, D. L. (2002), *A Biblical Theology of Exile,* Overtures to Biblical Theology, Minneapolis: Fortress Press, p. 94.

45 Joyce, P.M. (2007), *Ezekiel; A Commentary,* New York: T & T Clark, p. 3.

46 Joyce, P.M. (2007), *Ibid.,* p. 7.

47 Block, D.I. (2013), *By the River Chebar: History, Literary, and Theological Studies in the Book of Ezekiel,* Eugene: Cascade, p. 8.

This was questioned by multiple commentators starting in the early twentieth century. Hölscher proposed that only a seventh of the book originated from Ezekiel himself.[48] Torrey argued that the Book of Ezekiel is a pseudepigraph from the third century BC.[49] Using form-critical and traditio-historical analysis, scholars like Zimmerli, noted inconsistencies in the text and evidence of redaction.[50] He sees that significant elaborations to the book were made by a 'school of Ezekiel', particularly in relation to the oracles which express a message of an expectation of future salvation.[51] Redditt thought that anomalies in the sequence of the dates in the text pointed to subsequent redaction.[52] Scholars also have noted several layers within the text.[53] For example they discerned a priestly stratum in the book that is said to be different from Ezekiel's 'true' style.

Many commentators however, recognise the high degree of literary coherence in the book which is 'unmatched in the canon of biblical prophets'.[54] The literary unity is infused with a personal characterisation of Ezekiel suggesting that he is indeed the author of the work: (a) the prophecies in Ezekiel are consistently written in the first person, and read as though the prophet himself went through these experiences; (b) Several 'sign-acts' performed by Ezekiel are written in the first person. These *'memorabile'* are divine signs,[55] manifested

---

[48]   Hölscher, G. (1924), *Hesekiel, der Dichter und Das Buch,* BZAW 39, Giessen: Töpelmann.

[49]   Torrey, C.C. (1970), *Pseudo-Ezekiel and the Original Prophecy,* New York: Ktav Publishing.

[50]   Zimmerli, W. (1979), *Ezekiel 1,* Philadelphia: Fortress Press, p. 227. For example in verses Ez 1.4-28 there is evidence of several secondary expansions : (see section 'The Wheels as a Class of Angels').

[51]   Zimmerli, W. (1979), *Ibid.,* p. 62.

[52]   Redditt, P. L. (2008), *Introduction to the Prophets,* Grand Rapids: Eerdmans, p. 151.

[53]   Garscha, J. (1974), *Studien zum Ezechielbuch: Eine redaktionskritische Untersuchung von Ez 1–39,* Europäische Hochschulschriften 23, Bern: Herbert Lang.

[54]   Odell, S. (2005), *Ezekiel,* Smyth & Helwys Bible Commentary, Macon: Smyth & Helwys, p. 1.

[55]   Wolff, H. W. (1974), *A Commentary on the Prophet Hosea,* Philadelphia: Fortress, p. 10.

through personal actions of the prophet and in the book of Ezekiel there are strong indications that they were performed and written by the prophet himself; (c) Block says that the doleful words 'lamentations, moaning and woe,' *qînîm wāhegeh wāhî* (קִנִים וָהֶגֶה וָהִי) that are written on a scroll that Ezekiel is told to consume, (Ez 2:9- 3:3) are not just a metaphor for 'divine truths' but have a personal connection with Ezekiel's output,[56] and could somehow be key to what motivates Ezekiel. Greenberg, one of the foremost Ezekiel scholars, sees the text of Ezekiel holistically[57] and shows that careful analysis is able to reveal the personality of 'an individual mind'.[58] Block opines that the best analysis of ancient texts derives from those that have a positive and sympathetic disposition towards the author: looking for an understanding of complexity, rather than using it as grounds for cynicism.[59] Although it is likely that the book did undergo early redaction, it is generally accepted that the core of the book was probably written by the prophet himself.[60] It was likely essentially complete by the end of the sixth century BC. In summary, no coherent or sustainable critique has gained traction against the unity of the Book and the personal characterisation of Ezekiel that emerges from it.

Potential Environmental influences on Ezekiel's personality

When looking at Ezekiel's personality many commentators are struck by the florid, 'bizarre' features that abound in his writings, and this leads him to be labelled as 'strange'. He describes his visions in great detail and in addition is transported in them to Jerusalem and other places. There is also interaction

---

56 Block, D. I. (1997), *The Book of Ezekiel chapters 1-24*, Grand Rapids: Eerdmans, p. 20.

57 Levenson, J.D. (1984), 'Ezekiel in the Perspective of Two Commentators,' *Interpretation*, **38** 2, pp. 210–217; p. 213.

58 Greenberg, M. (1983), *Ezekiel 1-20*, New York: Doubleday, p. 26.

59 Block, D. I. (2013), *By the River Chebar: History, Literary, and Theological Studies in the Book of Ezekiel*, Eugene: Cascade, p. 257.

60 Clements, R. E. (1996), *Ezekiel*, Westminster Bible Companion, Louisville: Westminster John Knox Press, p. 6.

between his different visions, as the Chariot Vision is seen within two other visions,*[61] giving to Ezekiel a complexity of psychic experience.[62] The principal features which have been said to suggest an instability, illness or behaviour problem attributed to Ezekiel are:

Florid almost dreamlike visions with 'bizarre' features and hearing different associated noises. (Ez 1:18); 'Spiritual' travels. (Ez 3:12); texts delivered with apparent extreme anger and threatening disaster at times, (Ez 7:5-9); fixed extreme negative attitudes (towards Israel) (Ez 16; 22; 24); 'depressive features': 'fascination' with faeces and blood;[63] misogyny with a sexually debasing portrayal of women (Ez 23); long passages of repeated seemingly obsessive measurements with no clear purpose, (Ez 40–48); strange behaviours: making a hole in a wall, lying on his side for 390 days, told to use human faeces for cooking fuel (Ez 4:12), cutting his hair with a sword and tucking it into his garment; possible medical signs: mutism, or paralysis, (Ez 3:25–27; 4:4–8).

These different 'atypical features' have been used by different commentators to make medical diagnoses as follows:

1. Psychosis: has been suggested on the basis of a random aggregation of some of the above features, even suggesting schizophrenia.[64,65]

2. Schizophrenia: typically requires a combination of two of the following - delusions, hallucinations (usually auditory) and disorganized speech.[66]

---

[61] In Ez 10:6, the man clothed in linen from the Vision of the Executioners takes coals from the Chariot Vision. Also in Ez 43:2 the Chariot Vision enters the Vision of the Idealised Temple.

[62] Zimmerli, W. (1979), *Ezekiel 1*, Philadelphia: Fortress Press, p. 17.

[63] Block, D. I. (1997), *The Book of Ezekiel Chapters 1–24*, Grand Rapids: Eerdmans, p. 10.

[64] Van Nuys, K. (1953), 'Evaluating the Pathological in Prophetic Experience (particularly in Ezekiel),' *JBR* **21** (4), pp. 244–251.

[65] Poser, R. (2012), *Das Ezechielbuch als Trauma-Literatur*, Leiden: Brill, p. 14.

[66] American Psychiatric Association (2013) *Diagnostic and Statistical Manual of Mental Disorders,* 5th edn (DSM-5), Arlington: American Psychiatric Association. To meet the

The diagnosis has been suggested on the basis of a combination of 'catatonia' and 'paranoia' attributed to Ezekiel.[67] Ezekiel did not have disorganized speech and his experiences were visions, not hallucinations,[*68] with little in the way of auditory components (Ez 1:24). It is unfair to classify a prophet's statements as delusions,[*69] because belief in a prophecy is by definition, held only by the prophet, who is privileged to receive it. Ezekiel does not therefore have grounds for schizophrenia and psychosis is unlikely.

3. Personality Disorder. He is also said to have had a personality disorder.[70] For apparently abnormal behaviour to be diagnosed as a personality disorder it has to be 'an enduring pattern of behaviour that differs markedly from the person's culture'.[*71] Ezekiel's behaviour changes throughout the book. There is no single behaviour pattern that is persistent. Although his writing appears angry and threatening at times (Ez 7:5–9) at other times, particularly when replying to YHWH he is very meek, (Ez 4:14; 9:8). He does not therefore have an 'enduring pattern of behaviour'.

---

criteria for diagnosis of schizophrenia, the patient must have experienced at least 2 of the following symptoms: Delusions, Hallucinations, Disorganized speech.

[67] Lang, B. (1981), *Ezechiel: Der Prophet und das Buch (Erträge der Forschung)*, Darmstadt: Wissenschaftliche Buchgesellschaft, p. 61.

[68] Meta-awareness or deliberate attention to the content of the experience is retained in a vision, but not in a hallucination. See footnote 118 for discussion of meta-awareness.

[69] A delusion is defined as an idiosyncratic belief or impression maintained despite being contradicted by reality or rational argument.

[70] Broome, E.C. (1946), 'Ezekiel's abnormal personality,' *JBL*, **65**, pp. 277–92.

[71] Definition of a personality disorder: 'an enduring pattern of inner experience and behaviour that deviates markedly from the expectations of the individual's culture' taken from *https://www.nimh.nih.gov/health/statistics/personality-disorders.shtml*, accessed 14/02/2021.

4. Post Traumatic Stress Disorder (PTSD). Smith-Christopher sees Ezekiel's 'bizarre' behaviour as characteristic of a disorder such as PTSD.[72] PTSD can be brought on by a prolonged stressful situation, such as is seen in displaced refugees.[73] This author says in effect that it is impossible to assess Ezekiel fairly without weighing the effects of the extreme stresses he suffered. These include defeat and deportation of the Israelites, (the residents of Jerusalem looked upon exile as proof of divine rejection. Ez 11:15); the terrible experiences of war; his exile; loss of the priestly role he had been training for; the death of his wife. Smith-Christopher claims for example, that the tone of the Oracles of Judgement may have been heightened by the stress of adjusting to exile and loss of connection with his homeland.[74] He shows how passages from Lamentations (La 4:4–10) bring to life the harrowing situation in the siege of Jerusalem. Zimmerli also noted that the trauma of losing his wife may have had a psychological effect on Ezekiel.[75]

The diagnostic features of PTSD are tabulated in Appendix 10. The only two features that might fit with Ezekiel (which are highlighted) are one from group 3 (having angry outbursts) and one from group 4 (negative thoughts about oneself or the world). The diagnosis of PTSD would also require Ezekiel to have features from groups 1 and 2, and one other feature from groups 3 and 4. In addition, the frequency of PTSD

[72] Smith-Christopher, D. L. (2002), *A Biblical Theology of Exile*, (Overtures to Biblical Theology), Minneapolis: Fortress Press, p. 91.

[73] Lindert, J., et al. (2009), 'Depression and anxiety in labor migrants and refugees – a systematic review and meta-analysis,' *Soc Sci Med*, **69** (2), pp. 246–257.

[74] Smith-Christopher, D. L. (2002), *A Biblical Theology of Exile*, (Overtures to Biblical Theology), Minneapolis: Fortress Press, p. 81.

[75] Zimmerli, W. (1979), *Ezekiel 1*, Philadelphia: Fortress Press, p. 29.

in adults who have been refugees or experienced traumatic events is still only 25 to 31%.[76] The diagnosis therefore does not fit with Ezekiel.

5. Stroke. A recent suggestion about Ezekiel's mutism and lying on his side was that he had a stroke.[77] Although stroke can cause a prolonged, one-sided or bilateral weakness, if Ezekiel was mute from a stroke, it would have dramatically affected his ability to write, so this option is unlikely.

6. Conversion disorder. 'Hysteria' was suggested[78] as a cause for Ezekiel to lie on one side for a prolonged period (Ez 4:5). A conversion disorder is caused by an internal emotional stress that manifests as a physical weakness. A patient with a conversion disorder typically pays no attention to their disability, (symptom of *'belle indifference'*) rather than actually pointing out the cause of it as Ezekiel did. Also, the lying position he assumed was apparently voluntary and not secondary to any stated weakness.[79]

Divine Influences on Ezekiel's behaviour

The foregoing section agrees with commentators who have not supported suggestions of psychological disturbances or physical illness to explain Ezekiel's behaviour and writings.[80] Prophets are asked to do many unconventional

---

[76] Turrini G., et al. (2017), 'Common mental disorders in asylum seekers and refugees: umbrella review of prevalence and intervention studies,' *Int J Ment Health Syst*, **11**, (51).

[77] Mathew, S. K, Pandian, J. D. (2010), 'Newer insights to the neurological diseases among biblical characters of old testament,' *Ann Indian Acad Neurol*, **13**, pp. 164–166.

[78] Halperin D. J. (2007), In Rollins W. G., Kille D, A. eds, *Psychological Insight into the Bible: Texts and Readings*, Grand Rapids: Eerdmans, p. 123.

[79] We must also assume that Ezekiel actually did lie on one side for 390 days on his left side and 40 days on his right as commanded by the Lord, as this is not recorded.

[80] Childs, B. S. (1979), *Introduction to the Old Testament as Scripture*, Philadelphia: Fortress Press, p. 359.

actions*[81] and these do not necessarily reflect their true personality. Greenberg said that it is important to use a more reflective approach in discerning the character of the writer. Ezekiel is described as sensitive[82] and having 'an individual mind of powerful and passionate proclivities'.[83] Ezekiel brought about the death of Pelatiah the son of Benaiah, who died while Ezekiel was prophesying against him (Ez 11:13), which supports him as a powerful prophet. He gives the impression of one who was thoroughly prepared for spiritual leadership in the tradition of the priesthood.[84] Given that his trajectory to the priesthood was cut short at the age of 25 when he was exiled, he must have wondered what would become of this. The answer according to Block is that YHWH takes over his training, and he becomes a man 'totally possessed by the Spirit; called, equipped and gripped by the hand of God'.[85] This possession by the Spirit and gripping by the hand of God are likely what equipped a sensitive and passionate trainee priest to become a profound theologian and a powerful prophet – who was able to unapologetically carry out YHWH's commands to restore the stubborn rebels of the house of Israel.

There are five stages that can be discerned in Ezekiel's maturation as a priest/prophet:

1. Just as he reached the age of eligibility for priesthood, he is presented with one of the most vivid and complex visions described in the Bible in which he identifies the presence of YHWH (Ez 1:28). He must have understood from this that he had a privileged relationship with YHWH.

---

[81]   Isaiah was told to walk around naked for 3 years (Is 20:2). Jeremiah was told to hide his underwear under a rock (Je 13).

[82]   Block, D. I. (1997), *The Book of Ezekiel chapters 1–24*, Grand Rapids: Eerdmans, p. 11.

[83]   Greenberg, M. (1983), *Ezekiel 1–20*, New York: Doubleday, p. 26.

[84]   Block, D. I. (1997), *Ibid.*, p. 9.

[85]   Block, D. I. (1997), *Ibid.*, p. 11.

2. His commissioning is next, with extreme descriptions of the rebelliousness and stubbornness of Israel.*[86] If Ezekiel did not know it from having been exiled himself, he is told how deeply the people have offended YHWH (Ez 2:3). Ezekiel is assured that YHWH will 'make his forehead like emery', *kešāmîr ḥāzāq miṣṣōr* (כְּשָׁמִיר חָזָק מִצֹּר) which was the hardest stone know at the time, harder than flint.[87] If YHWH is giving Ezekiel extreme comparisons like this, it is not surprising that the language of his oracles is harsh.

3. He is told to eat the scroll with the words 'lamentations, moaning and woe' on them. Block says that these words describe the effects on the people of the judgements Ezekiel will pronounce.[88] However they could also describe the situation in Jerusalem at the time of the fall,[89] and that Ezekiel's writings were YHWH's loving response to that deep misery that Israel brought upon itself. The words could therefore be strongly motivating for Ezekiel.

4. The subsequent command for Ezekiel to lie on his left side for 390 days and his right side for 40 days to bear the iniquity of the houses of Israel and Judah, (Ez 4:4-6) makes Ezekiel 'a suffering servant'.*[90],[91] This period of enforced lying was likely a 'desert' or 'wilderness' experience for Ezekiel that would become traditional training for spiritual growth. Additionally, all the sign-acts (Ez 4:1- 5:4) have a common feature of

---

[86] Ezekiel later compares their insensitivity with 'a heart of stone' (Ez 36:26).

[87] Block, D. I. (1997), *The Book of Ezekiel chapters 1–24*, Grand Rapids: Eerdmans, p. 129.

[88] Block, D. I. (1997), *Ibid.*, p. 125.

[89] Smith-Christopher D. L. (2002), *A Biblical Theology of Exile*, (Overtures to Biblical Theology), Minneapolis: Fortress Press, p. 59.

[90] Rashi says that affliction is brought on Ezekiel in order to wipe away the iniquities of Israel (Babylonian Talmud 39a).

[91] Block, D. I. (1997), *Ibid.*, p. 177.

afflicting the prophet.[92,93] 'The event which is proclaimed by the prophet seizes him again and again and makes him a part of the event itself.'[94]

5. Finally, Ezekiel sees the visions of the abominations in the Temple and the pitiless slaughter in the Vision of the Executioners. YHWH's requirement for justice in the face of the great guilt of Israel and Judah justifies: 'my eye will not spare, nor will I have pity'*[95] and helps to explain Ezekiel's extreme language and excoriating oracles.

Did Ezekiel turn to writing rather than proclaiming?

Recently there has been strong support that the book of Ezekiel was *written* by Ezekiel himself.[96,97] The cohesive structure and distinctive tone of the writings and oracles made early commentators conclude that the book was a written document by a single author.[98] Factors that supported this are, that the construction of the book around the recurring Chariot Vision hold the book together and that careful chronological dating of many sections of the book suggest a diary type of written record. Towards the late nineteenth century, new methods of critique were applied to the book which suggested that the book was in fact reworked by Ezekiel or a school of followers,[99] and that it was

---

92 Origen (2010), *Homilies 1-14 on Ezekiel*, trans Sheck, T. P., New York: Newman Press, p. 32. (Hom, in Ezech. 1:4.1-2) said 'If you wish to hear Ezekiel, the son of man, preaching in captivity, understand him as a type of Christ.'

93 Greenberg, M. (1983), *Ezekiel 1-20*, New York: Doubleday, p. 127.

94 Zimmerli, W. (1964), *VT,* **XV** (33), pp. 515-527; p. 517.

95 Ezekiel is told to say 'Thus says the Lord God:...I will do to you what I have never done before and will never do again...I will not look on you with pity or spare you' (Ez 5:9-11).

96 Greenberg, M. (1983), *Ezekiel 1-20*, New York: Doubleday, p. 26.

97 Davis, E. F. (1989), *Swallowing the Scroll: Textuality and the Dynamics of Discourse in Ezekiel's Prophecy*, Bible and Literature Series 21, Sheffield: Almond, p. 19.

98 Odell, S. (2005), *Ezekiel*, Smyth & Helwys Bible Commentary, Macon: Smyth & Helwys, p. 2.

99 Zimmerli, W. (1979), *Ezekiel 1*, Philadelphia: Fortress Press, p. 62.

more likely to be a record of a collection of speeches. However, recently, elements of the book that suggest a written text have been reaffirmed.[100] Features that suggest a written record include:

1. The fact that the book was written in the first person with YHWH speaking to Ezekiel and that Ezekiel is personally told by YHWH to perform several specific actions and the prophet relates what he does.[101]

2. The style of writing: Davis says that Ezekiel is resourceful in directing the audience's attention and understanding and punctuates his account with markers that make reading and understanding easier. Some of the formal features of the book appear to function in this way.[102] Examples of these are: 'I am YHWH, I have spoken' which appears 14 times;[103] 'set your face towards...' occurs 9 times.[104]

3. Block points out that Ezekiel has a distinctive literary style that is recognisable by: i) the prominence of extended narrative presentations; ii) frequent use of parallelism in his oracles; iii) frequent use of Aramaisms with twice as many as his contemporary Jeremiah, and frequent use of Accadianisms; iv) he has 126 hapax legomena (Jeremiah has only 88) many of which reflect his Babylonian surroundings.[105]

4. Writing may be a way of ordering experience and rendering it communicable. A written work could have allowed Ezekiel more complex

---

[100]   Odell, S. (2005), *Ezekiel,* Smyth & Helwys Bible Commentary, Macon: Smyth & Helwys, p. 3.

[101]   Wilson, R.R. (2000), 'Ezekiel,' in *Harper Collins Bible Commentary,* ed. Mays J. L. *et al.,* San Francisco: Harper San Francisco, p. 588.

[102]   Davis, E. F. (1989), *Ibid.,* p. 77.

[103]   Block, D. I. (1997), *The Book of Ezekiel Chapters 1–24.* Grand Rapids: Eerdmans, p. 37.

[104]   Block, D. I. (1997), *Ibid.,* p. 34.

[105]   Block, D. I. (1997), *The Book of Ezekiel Chapters 1–24,* Grand Rapids: Eerdmans, p. 40.

interaction between the writer and reader and a deeper level of critique, than would have been possible with verbally delivered criticism.[106] Wilson says that Ezekiel may have turned to writing as a way of expressing his views because his harsh castigation of the people may have made them increasingly hostile, causing him to curtail his public activities.[107] An interesting process that is possible with a written rather than a wisdom tradition is a theological reflective ability which allows the writer to systematize and project past prophecies onto future events,[108] (e.g. Ez 38:17).*[109] In this way, cultural and religious symbols can be reformulated to lay the foundation for the future.[110]

Conclusion of chapter

There is no medically cogent support that Ezekiel had any mental or physical illness. The power and unconventional nature of his writings are likely due to the response of his passionate nature to the extreme situation he and his people were in and the vividness and spiritual content of his visions and prophetic revelations. Personal details and the precise chronology recorded in the book make it likely the book was written personally by Ezekiel. The significance for the thesis is that the text is unique in the Old Testament as a first-hand, written account of profound mystic experience. It demands in depth analysis in its

---

[106]   Davis, E. F. (1989), *Swallowing the Scroll: Textuality and the Dynamics of Discourse in Ezekiel's Prophecy,* Bible and Literature Series 21, Sheffield: Almond, p. 19.

[107]   Wilson R. R. (1980), *Prophecy and Society in Ancient Israel,* Minneapolis: Fortress Press, p. 285.

[108]   Odell, S. (2005), *Ezekiel,* Smyth & Helwys Bible Commentary, Macon: Smyth & Helwys, p. 465.

[109]   Ez 38:17. 'Thus says the Lord God: Are you he of whom I spoke in former days that by my servants the prophets of Israel, who in those days prophesied for years that I would bring you against them?'

[110]   Davis, E. F. (1989), *Ibid.,* p. 106.

original language laying aside speculative concerns about Ezekiel that would diminish its power and veracity.

Chapter 2

Chariot Vision Background

Overview of chapter

This chapter first looks at the different possible ontological / existential natures of the Chariot Vision: ontologically, whether the vision represents God, an intermediary (or *shekhinah*) or an analogy of an aspect of God; existentially, whether it is an objective entity or a subjective construct in or of Ezekiel's mind. Secondly ten different possible theological meanings of the vision are reviewed. The significance of the Chariot Vision for the Book of Ezekiel is reviewed, its appearance within other visions and its links with other visions. The main significant Chariot Vision interpretations of seven Patristic authors and nine modern authors, as well as ancient and medieval Jewish interpretations, are reviewed.

Introduction

The initial manifestation and description of the *Merkavah* or Chariot Vision is unique. It intrigues the reader through its complex detail, its hierarchical structure and its brisk movements, but particularly because it presents to us the 'Glory of God', the appearance of which causes Ezekiel to fall down to the ground. The vision also represents Ezekiel's baptism into his prophetic life. But the vision is also a dramatic prelude to the mysteries that await the reader.

What was the Ontological/Existential Nature of the Chariot Vision?

There are three main possibilities for the ontological nature of the Vision:

It is a vision of God himself or God and his 'chariot'

The vision is unlikely to be a direct vision of God himself because the figure on the Chariot of YHWH is divided at the waist.[111] Additionally, the vision is complex with mystical four-faced figures (the 'Living Creatures') that are unlikely to be existing spiritual beings. The representation of the wheels with multiple eyes on them, also are unlikely to be existing spiritual entities. On the other hand, towards the end of the initial Chariot Vision, Ezekiel states that when he saw the vision he 'fell upon his face' and he heard a voice, (Ez 1:28). So even if this was not a vision of God himself, Ezekiel felt he was in the presence of the Lord, and shortly later he saw a hand stretched out to him, he heard a voice telling him to eat the scroll in it, and he ate it. This confirms the close presence of the Lord with this vision.

It is a vision of an intermediary between God and man (like the shekhinah or metatron)[112]

Ezekiel describes the vision as 'the likeness of the Glory of the Lord' (Ez 3:28). Biggs says that the 'Glory of the Lord' is a technical term for the presence of God among the people of Israel.[113] The sign of its presence was a cloud (Ex

---

[111] Fisch, S. (1950), *Ezekiel,* London: The Soncino Press, p. 8.

[112] Rabbinic scholars put forward intermediaries through which God was able to contact the world: wisdom, metatron and *shekhinah*. *Shekhinah* (שכינה) is the Hebrew word for 'dwelling' or 'settling' and is said to denote the divine presence of God on earth. *Metatron* (מטטרון) is derived from the Latin *mētātor*: 'one who metes out or marks off a place,' 'a measurer'.

[113] Biggs, C. R. (1996), *The Book of Ezekiel,* London: Epworth Press, p. 5.

40:35). The cloud in Exodus thus represented the actual presence of the Lord. In Jewish commentary, the 'Glory of God' of the Chariot Vision was identified with the *Shekhinah*.[114] The fact that Ezekiel calls the Chariot Vision the 'glory' of God suggests that he identifies an attribute of God with the Vision. If *'Shekhinah'* denotes the closeness of God, and the Vision represents that which is not God himself, but denotes an actual 'presence' of God, then the Vision could be considered as a *'Shekinhah'* or an intermediary between God and the world, that dispense the power and will of God, (if such intermediaries exist). One has therefore to differentiate whether the Chariot Vision was actually an appearance of a real intermediary rather than a vision. The Chariot Vision appears within other visions: that of the executioners (Ez 10) and that of the idealised temple (Ez 43). There is an interaction between two visions in that the visionary man clothed in linen (in the Vision of the Executioners) takes fire from the Chariot Vision and introduces it into the Vision of the Executioners (Ez 10:7). This makes it unlikely that the Chariot Vision is an actual appearance of a substantial intermediary like the *shekhinah*.

It is a vision of an analogy of an aspect of God, or of God's immanence

Is the Chariot Vision a vision of an 'analogy'? This would imply that the actual chariot structure does not exist outside of the Chariot Vision but is a structure that conveys particular revelatory information for the time and *Sitz im Leben* of Ezekiel and constitutes part of his prophetic message. There are Mesopotamian features in the vision, particularly in the four-faced 'Living Creatures'. The 'Living Creatures' have several links to ancient Mesopotamian art and statuary (particularly the Guardians of Nimrud or *Lamassu*[*115]) – this is explored in detail below. These features do support the Vision as an analogy.

---

[114] Scholem, G. (1961), *Major Trends in Jewish Mysticism*, New York: Schocken, p. 66.

[115] The Lamassu was first depicted as a symbol of power in about 967 BC when Tiglath-Pileser II came to the throne. See subsection 'The Lamassu'.

If the Chariot Vision is an analogy, firstly, was it an objective vision or a subjective mental experience of Ezekiel? For example, were the fire and *hašmal* actually present in the storm cloud (Ez 1:4), or was it all a subjective mental phenomenon? Secondly, is it a divinely initiated objective construct that reflects real attributes of God that Ezekiel was privileged to see; or is it a subjective dream-like synthesis of Ezekiel's mind of his understanding of the glory of God?

The character of the vision is often quite dreamlike and Ezekiel uses several words to qualify the vision, for example: 'likeness' *děmûṯ* (דְּמוּת) in the description of the 'appearance of a man' (Ez 1:26).[116] 'The likeness' is used multiple times in the description of the Chariot Vision as well as 'something that looked like' for the coals (Ez 1:13), or the 'appearance of' for the wheels (Ez 1:16). Even if the vision was objective, Ezekiel expresses uncertainty about what he is actually seeing.

Ezekiel specifically gives a geographic landmark (the River Chebar: Ez 1:3)*[117] and the first vision starts in a storm cloud. This strongly suggests this first vision was objective. Ezekiel also says, 'the heavens were opened' (Ez 1:1), suggesting that the vision was seen at least initially as objective. Would someone else standing near Ezekiel have seen it as well? Probably not, as in Ez 8:1, Ezekiel states that he sat in his house with the elders of Judah before him and 'the hand of the Lord God fell there upon me'*[118] and he then had a vision of 'the Glory

---

[116] Block, D. I. (1997), *The Book of Ezekiel Chapters 1–24*, Grand Rapids: Eerdmans, p. 107.

[117] The River Chebar has been identified as a branch of an elaborate canal system joining the Tigris and Euphrates, in the vicinity of Nippur. Block, D. I. (1997), *The Book of Ezekiel Chapters 1–24*, Grand Rapids: Eerdmans, p. 84.

[118] Modern neuroimaging research on trance, shows that it is an 'absorptive' state, meaning that the individual's train of thought is internally directed and disengaged from the surrounding sensory environment. The 'shamanic' trance (a volitional, self-induced state of altered consciousness) is a goal-directed state that explores a pre-specified problem. The practitioner experiences an inner reality filled with imagery, spirits and symbols and the trance is often experienced as a lucid dream with meta-awareness (a state of deliberate attention toward the contents of conscious thought, serving as an appraisal of experiential

of the God of Israel', 'like the vision he saw on the plain' (Ez 8:4). There was no indication that the elders saw the vision too. This is similar to St. Paul's vision (Ac 9:7), where the onlookers did not see his vision, but did hear what Paul was told by the Lord. Determination of whether the vision was objective or subjective may therefore not be a valid distinguishing criterion since even objective visions seen by some are not seen by everyone. The Chariot Vision in Ez 1:4 is more likely a 'privileged' objective phenomenon. The Chariot Vision is also described by Ezekiel as interacting with the Vision of the Executioners in Jerusalem (Ez 10: 6-8). The Vision of the Executioners was not a real objective event however as there is no record from Jeremiah that anything like this happened. In these latter two instances, the Chariot Vision was probably more a subjective vision rather than objective. There is also no evidence that Ezekiel was ever transported physically to different places in his visions. To help differentiate whether the vision was a subjective personal synthesis or dreamlike state, or a divinely initiated construct, I will look at different explanations for the Chariot of YHWH and the elements of the Vision.

Sweeney sees the Chariot Vision as the divine presence.[119] He sees an analogy to the Ark of the Covenant which was housed in the holy of holies or the inner

---

consciousness. It is not only having an experience but *knowing* that one is having an experience), and insight. (Hove, M. J. *et al.,* (2016), 'Brain Network Reconfiguration and Perceptual Decoupling During an Absorptive State of Consciousness,' *Cerebral Cortex,* **26** (7), pp. 3116–3124). Dreaming in sleep parallels shamanic trance as a visually dominated exploratory state with high posterior corpus callosum activation, but conversely, it is a haphazard 'exploratory' state beyond one's control. Dreaming is associated with decreased activity in certain brain regions (lateral prefrontal cortex) involved in cognitive control and metacognition, (which is awareness and understanding of one's own thought processes), with a corresponding lack of insight and meta-awareness. (Fox, K.C.R. (2013) *Front Hum Neurosci,* **7**, article 412). Lucid dreaming, where a person maintains awareness and reflective consciousness is not uncommon, however. The implication of this for Ezekiel is that his episodes where the hand of YHWH came upon him are likely trances, not dream-like states, because his ability to describe them vividly and write them down suggests preserved insight and meta-awareness. The goal-directness of trances tends to support a pre-specification of the content of the vision by the spirit.

[119] Sweeney, M. A. (2013), *Reading Ezekiel,* Macon: Smyth & Helwys, p. 26.

chamber of the Jerusalem Temple until its destruction in 587 BC.[120] He sees the Chariot Vision as portraying the Ark as YHWH's throne-chariot that conveys YHWH through the heavens. He sees the imagery of the Vision as a whirlwind with thick cloud - which recalls the Jerusalem Temple in times of worship, filled with incense smoke and menorahs or lamp stands burning to symbolise YHWH's presence. The crystal firmament below the throne is the counterpart of the gold of the mercy seat of the Ark (Ex 2:17). He says that the multiple eyes on the wheels could be interpreted as the 'eye' of the divine presence looking throughout the world. However, he thinks they could also just be the reflection of light from the wheel's polished surfaces.[121] The Ark was conveyed to Jerusalem on a wheeled cart in the reign of King David (2Sa 6:3[122]). There is also imagery in the Psalms (Ps 68:17[123]) and in Habakuk (Hb 3.8[124]) of God using a chariot. Sweeney feels that the Chariot Vision was a composite from Ezekiel's memory. The Ark was however laid on a cart not a chariot and the imagery of the Chariot Vision is far more complex than the Ark of the covenant, so this does not explain the whole vision. Vogt sees a parallel with the temple laver which had four wheels (1Ki 7:30).[125] Greenberg[126] says that although there were elements in the vision that could have been drawn from Israelite tradition and surrounding iconography, the vision was so novel that he excludes the possibility that it was drawn out of Ezekiel's memory. He says

---

[120] Sweeney, M. A. (2013), *Ibid.*, p. 29.

[121] Sweeney, M. A. (2013) *Ibid.*, p. 65. The wheels had the appearance of *taršíš* (Ez 1:16) which is a precious stone that Block says equates with Spanish gold topaz and might have been reflective. Block, D. I. (1997), *The Book of Ezekiel Chapters 1–24*, Grand Rapids: Eerdmans, p. 99.

[122] 2 Sa 6:3. 'they carried the ark on a new cart' *agalah* (עֲגָלָה) or cart is used here not *rechuv* (רכוב) which is chariot.

[123] Ps 68:17. 'The chariots of God are myriads, even thousands upon thousands'

[124] Hb 3:8. 'when you drove your horses, your chariots to victory?'

[125] Vogt, E. (1981), *Untersuchungen zum Buch Ezechiel*, Analetica Biblica 95, Rome: Pontifical Biblical Institute Press, p. 69.

[126] Greenberg, M. (1983), *Ezekiel 1–20*, New York: Doubleday, p. 58.

that Ezekiel believed the Chariot Vision to be the 'majesty of God' and this is why he fell awestruck to the ground.[127] However, it was not so much a vision of God, as a vision of 'divine heavenly realities'[128] as the first verse of Ezekiel (Ez 1:1) says 'the heavens opened and I saw visions of God'.

The text of Ezekiel gives some insights into the Chariot Vision. Firstly, compared with the Idealised Temple Vision (Ez 40 to 48), for example, the Chariot Vision is more complex. It has moving components, makes noise as it moves and it contains burning coals. Its actions, particularly in relation to exiting and returning to the Temple are central to the whole book of Ezekiel. The fact that the Chariot of YHWH left the temple by the East Gate (Ez 10:18) after the Vision of the Executioners, and the desecration of the Temple,[129] supports that the Chariot is indicative of the divine presence. In the ancient Near East, it was believed that no temple was destroyed unless its god had abandoned it.[130] Its return Ez 43:1 from the east to the new temple means the 'divine patron of the nation is returning'.[131]

In summary, apart from the initial Chariot Vision, Ezekiel's visions appear to be internal subjective experiences. However, the vision did convey to Ezekiel that he was in the presence of God, making the vision more than just an analogy. In his later visions, his transportations appear to be purely mental, with Ezekiel being in prophetic ecstasy[132] (or altered state of consciousness experience*[118]). In Ez 40:2 he says 'the hand of the Lord was upon me and brought me *in the visions of God* into the land of Israel and set me down upon a very high

---

[127]   Greenberg, M. (1983), *Ezekiel 1–20*, New York: Doubleday, p. 52.

[128]   Block, D. I. (2013), *By the River Chebar: History, Literary, and Theological Studies in the Book of Ezekiel*, Eugene: Cascade, p.150.

[129]   Block, D. I. (1997), *The Book of Ezekiel Chapters 1–24*, Grand Rapids: Eerdmans, p. 326.

[130]   Pritchard, J. B. (ed.) (1969), *Ancient Near Eastern Texts Relating to the Old Testament*, 3rd edn,. Princeton: Princeton University Press, p. 455 ff.

[131]   Block, D. I. (1998), *The Book of Ezekiel Chapters 25–48*, Grand Rapids: Eerdmans, p. 578.

[132]   Poser, R. (2012), *Das Ezechielsbuch als Trauma-Literatur*, Leiden: Brill, p. 33.

mountain..."*[133] This vision is located on a high mountain in Israel, but the geographic description is not precise as it is in the initial Chariot Vision. In conclusion therefore, it seems most likely that despite the presence of elements of contemporary Mesopotamian tradition, the vision was an objective, divinely initiated construct and not a synthesis of Ezekiel's mind. It is likely therefore that the Chariot Vision is an objective construct that reflects real attributes of God, or denotes the presence of God.

The Inaugural Vision: Theological significance

The theological significance of this inaugural vision is profound. Odell says that the Vision is a representation, of the form of the divine glory and not the divine glory itself. She says that it is a highly stylized meditation on God's commanding presence in the world.[134]

The inaugural vision (Ez 1:4-28) can be divided into five sections: The introduction (Ez 1:1–3); The Living Creatures (Ez 1:4–14); The wheels (Ez 1:15–21); The platform and the throne (Ez 1:22–28) and Conclusion.[135] Block says that several aspects of proclamation are present in the Chariot Vision:

1. The vision proclaims the transcendent glory of God.[136] The Hebrew word *nōgah* (נֹגַהּ) (brilliance or effulgence) is used in Ez 1:28 to describe the brilliance of the Glory of God. The term is used to express the combination of presence with transcendence of YHWH.[137] *Nōgah* is also used to describe the power of God in the prayer of Habakuk, when

---

[133] Joyce argues that this narrative could be evidence of a heavenly ascent by Ezekiel. Joyce, P. M. (2007), 'Ezekiel 40–42: The Earliest "Heavenly Ascent" Narrative?' in de Jong H. J., Tromp, J. (eds), *The Book of Ezekiel and its Influence*, Aldershot: Ashgate, p. 24.

[134] Odell S. (2005), *Ezekiel*, Smyth & Helwys Bible Commentary, Macon: Smyth & Helwys, p. 32.

[135] Block, D. I. (1997), *The Book of Ezekiel Chapters 1–24*, Grand Rapids: Eerdmans, p. 91.

[136] Block, D.I.(1997), *Ibid.*, p. 106.

[137] Blenkinsopp J. (1990), 'Ezekiel,' *Interpretation: A Journal of Bible and Theology*, John Knox Press: Louisville, p. 19.

YHWH comes with glory and retribution to measure the earth (Hb 3:4).[138] The Glory of God is apparent in several other biblical theophanies, without explicit mention of 'brilliance'. (see Appendix 4: Ex 24:17; Re 4:8)

2. The Chariot Vision also proclaims the holiness of YHWH. There is a similarity with Isaiah's calling, where angels had three pairs of wings (Is 6:2) and with Revelation where the Living Creatures also had three pairs of wings and in both of which they cried out 'Holy! Holy! Holy!' (Re 4:8). In Ezekiel the angels had a second pair of wings with which they covered their bodies, like the angels in Isaiah. Block says covering their bodies with their wings distinguishes the created beings in the Chariot Vision from YHWH.[139]

3. The sovereignty of YHWH is proclaimed in the vision. Launderville[140] says that the Chariot Vision is a spiritualization or interiorization of the traditional model of royal authority that was the cherubim throne in the Temple in Jerusalem (1Ki 8:7). He sees the Chariot Vision as a transformation of the cherubim throne into the mental model of a throne-chariot. The Chariot signifies that order in the world arises from YHWH's sovereign governance. In the Chariot Vision, Ezekiel makes an abstract point about the workings of power in the heavenly sphere. The Chariot was critical for guiding Ezekiel and the exiles to understand how YHWH's sovereign power governed their lives even after the Davidic king and Jerusalem were lost to them.

4. The vision also shows God's identity with his people. God wonderfully makes an appearance in a likeness of the human form (Ez 1:26). This

---

138 Hb 3:4 'His brightness was like the light.'

139 Block, D. I. (1997), *Ibid.*, 107.

140 Launderville, D. (2004), 'Ezekiel's Throne-Chariot Vision: Spiritualizing the Model of Divine Royal Rule,' *Catholic Biblical Quarterly*, **66**, pp. 361–377; p. 361.

is not an actual representation of God but a 'reflection' of God's inner human form, declared in Ge 1:26: 'let us make man in our own image'. For the first time in the scriptures, he presents a visual image of a human-like form publicly to us through Ezekiel's vision. This revelation is deeply consolatory, as it shows God's closeness to his people and that in the time of trial that is the Exile, he shows himself as one of us.

5. In the Chariot Vision, YHWH reveals a manifestation of his Glory to his people who are not living within his city of Jerusalem or within his land. This not only confirms Ezekiel in his priesthood-without-a-temple, but also shows that the Lord, in all his glory, is manifest wherever in the world his people are and not just in the Temple in Jerusalem. This must have been reassuring to the exiles who had been accused of being rejected (Ez 11:17). It also hints at a new way of adoring God, which is not tied to location.[141] (see subsection: 'The Christocentric Interpretation')

6. There are also echoes of impending judgement in the Chariot Vision. Block mentions the burning coals between the wheels in the initial Chariot Vision which are later spread over Jerusalem after the Executioners have ruthlessly passed through the city, killing those who defiled it (Ez 10:2). The Glory of God presented in Ezekiel's first vision leaves the Temple by the East Gate after the Temple is defiled but returns without warning to the reconstructed temple in Ez 43:2. This recalls Malachi (Ma 3:1) who prophesies that the Lord will suddenly enter his temple on the day of his coming for judgement.[142] The wheels of the chariot can be seen as representing justice.

---

[141] Smith J. Z. (1987), *To Take Place: Towards Theory in Ritual*, Chicago: University of Chicago Press, p. 117.

[142] Malachi, prophesied after Ezekiel in about 420 BC.

7. The vision shows Ezekiel a clear vision, that his ministry is a vocation like no other. It represents conscription into the service of the King of Kings and Lord of Lords who sits on his glorious throne unrivalled in majesty and power.

Three further aspects of the Chariot Vision can be added to those of Block:

1. The Chariot Vision is a sign of righteousness.*[143] Block sees the descent of the Glory of God' in the Chariot Vision as foreshadowing the descent of Christ to dwell among us, 'full of grace of truth' at the Incarnation.[144] The initial Chariot Vision carries with it the promise of restoration and of resurrection that in the right place, time and season will be fulfilled. Although the Glory of God leaves the Temple at the time after its desecration it returns at a future time to a perfected temple. The Idealised Temple Vision is a lofty spiritual ideal showing both God's perfection and the perfection of his people, with individual Christians becoming temples and residences of God.[145]

2. The Chariot structure can be interpreted as an analogy and insight into how God reveals himself to the world[146]. It is argued in this thesis that a rational science-based hierarchy is present in the structure and movements of the Chariot and implies that 'science' is a way in which God communicates with man. In contrast to the theophany in Isaiah (Is 6:5-7), where there is a tension between uncleanness and holiness; in the Chariot Vision the tension is between creator and creature,[147] (Ez

---

[143] Righteousness is the perfection of God himself and also describes the state of man justified through the Redemption. Connaughton, L. (1980), *A–Z of the Catholic Church*, Leigh-on-Sea: Kevin Mayhew, p. 167.

[144] Block, D. I. (1998), *The Book of Ezekiel Chapters 25–48*, Grand Rapids: Eerdmans, p. 746.

[145] Block, D. I. (1998), *Ibid.*, p. 506.

[146] See chapter 'Revelation, Rationality and Science'.

[147] Zimmerli W. (1979), *Ezekiel 1*, Philadelphia: Fortress Press, p. 101.

1:28; 2:1). This is in keeping with the hierarchy in the Chariot Structure with the transcendent glory of the 'likeness of a man' image at the top and the wheels which are a human invention, at the bottom.

3. It is argued in this thesis that the Chariot Vision enfolds a synthesis of reason and spirit. Ezekiel is the prophet of the spirit,[148] and the power of the spirit is central to the Chariot Vision and its movements. As noted in the preceding subsection 9, the chariot is structured in a rational way. This structure can be shown to reflect different spheres of interaction between reason and spirit and the interaction at different chariot levels appear to have distinctive revelatory significance.

Significance for the book as a whole and interaction with other visions.

The Chariot Vision and its 'Glory of the Lord' theme (Ez 1:28) are a central part of the structural fabric of the Book of Ezekiel. The reasons for this are: (a) the vision is described in great detail in the first few verses even before Ezekiel is called to be a prophet.(b) The vision is seen repeatedly throughout the first eleven chapters of the book and again in Ez 43 (Ez 1:4; 3:23; 9:3; 10:1-22; 11.23; 43:2-4); also, in Ez 8:2 the 'likeness of a man' is seen alone. In Ez 44:4, the 'Glory of God' is seen without the chariot. The vision is most fully described in chapter 1, but further details are given in Ez 10.1. The second time that Ezekiel sees the Chariot Vision is about 13 months after the first. The description of the second of these visions is more concrete and less analogical. Block says that the reason for this is that Ezekiel is now more familiar with the vision and he is able to describe the vision in a rational, composed and coherent fashion.[149] (c) The vision is incorporated into the Vision of the Executioners in Ez 8. (d)

---

[148] Block, D. I. (2013), *By the River Chebar: History, Literary, and Theological Studies in the Book of Ezekiel*, Eugene: Cascade, p. 142.

[149] Block, D. I. (1997), *The Book of Ezekiel Chapters 1–24*, Grand Rapids: Eerdmans, p. 90.

Twenty years after the initial vision, the Chariot Vision is seen for the last time within the Idealised Temple Vision in Ez 43.

Links with other visions.

The Spirit of YHWH that is in the Living Creatures and the wheels, links the Chariot Vision with the Vision of the Dry Bones (Ez 37:1-14): it causes both the movement of the chariot in the Chariot Vision (Ez 1:12) as well as the movement of the bones as they are clothed in flesh and move, in the Vision of the Dry Bones (Ez 37:10). The spirit therefore links these two visions.

It is argued in this thesis that the wheels of the Chariot Vision have a metrological function and denote measured movement of the Chariot over the earth (see 'Lower Tier', Discussion). Measurement is central to the Idealised Temple Vision and the Vision of Water Flowing from the Temple and is a central theme throughout the Book of Ezekiel.

Eschatology in Ezekiel and in the Chariot Vision.

As mentioned above, Ezekiel, with Isaiah (Is 24), is one of the main early eschatological prophets. The prophecies against Gog and Magog (Ez 38; 39) are particularly known for their eschatological message,[150] and have been linked to their mention in Re 20:7–10 in the battle against Satan at the eschaton. Also, many features in the reconstructed idealised temple section (Ez 40–48) commend an eschatological interpretation,[151] as seen in the divine speech (Ez 43:6–12). Odell says that there are 3 parts in the divine speech.[152] In the first part, YHWH declares that the new temple is the place of his throne and foot-

---

[150]   Block, D. I. (2014), *Beyond the River Chebar: Studies in Kingship and Eschatology in the Book of Ezekiel*, Cambridge: James Clarke & Co., p. 95.

[151]   Block, D. I. (1998), *The Book of Ezekiel Chapters 25–48*, Grand Rapids: Eerdmans, p. 504.

[152]   Odell S. (2005), *Ezekiel*, Smyth & Helwys Bible Commentary, Macon: Smyth & Helwys. p. 497.

ftool. This means that the temple is a perfect merging of the two realms of heaven and earth.*[153] Second, YHWH identifies the central abuse leading to the abandonment of the Temple and demands that the house of Israel puts an end to these practices (Ez 43:7–9). The new temple is now completely free of any activity associated with idolatry. Thirdly, YHWH tells Ezekiel to declare the dimensions and ordinances to the house of Israel and to record them in writing. The intended effect of the vision is 'so that they may be humiliated'.[154] YHWH challenges the people to consider their role in the desecration of the divine name by their iniquitous behaviour. The Idealised Temple Vision presents a spiritual map of perfect holiness that puts them to shame. They are invited into it by God's good grace. If YHWH enters by the East Gate, shuts the door behind him and takes up permanent residence, then Ez 16:62–63[155] is indeed fulfilled.[156] The fact that the Idealised Temple Vision is intended to put forward God's perfect holiness, to proclaim the mercy of God and expose the sinfulness of human beings,[157] gives it an eschatological character. The spiritual ideal as expressed in this temple is in reality, a target that mankind can never achieve, but will be achieved in the elect at the eschaton. The marking of the foreheads with the sign of the Tau (תו) in the Vision of the Executioners (Ez 9:4–6) is another eschatological link as it is reminiscent of the sealing of the foreheads of the servants of God in Revelation (Re 7:3).

Within the Chariot Vision, Gregory the Great sees an eschatological meaning in the four Living Creatures, because they emerge from a fiery cloud with

---

[153] Similar to merging of realms in the Vision of Water Flowing from the Temple.

[154] Block, D. I. (1998), *Ibid.*, p. 589.

[155] 'Then I will establish my covenant with you and you will know that I am YHWH, in order that you may remember and be ashamed, and that your opening of the mouth will never happen again because of your disgrace once I have purified you of all that you have done. The declaration of the Lord YHWH' (Ez 16:62–63).

[156] Odell S. (2005), *Ezekiel*, Smyth & Helwys Bible Commentary, Macon: Smyth & Helwys, p. 500.

[157] Block, D. I. (1998), *Ibid.*, p. 590.

brightness.*[158] Gregory concludes that the 'cherubim' in the Chariot Vision (Ez 10:7) represent the saints who have all been made perfect through the gospels and who return with Christ at the end of time.[159] A further feature of the Living Creatures that suggests an eschatological meaning is that the four faces can be seen to represent the four quadrants of the heavens (see 'Middle Tier'). This gives a timeless and cosmic character to the Chariot Vision.

Block feels that Ezekiel presents a spiritual ideal rather than an eschatological one.[160] He says that although there are suggestions of eschatological features, there is no accompanying eschatological language. Instead, he feels the Idealised Temple Vision presents a lofty spiritual ideal in a theological constitution for a new Israel: a foundation for a spiritualization of the temple that extends to individuals, as 'where God is, there is order and the fulfilment of all his promises'.

Church Fathers' interpretation of the Chariot Vision

Irenaeus of Lyons (c.130-c.200)

A connection between the four creatures of Ezekiel and the four gospels was initially conceived by Irenaeus of Lyons. Irenaeus starts from an understanding that the Logos/Word is active in the Old Testament as well as in the New Testament. Irenaeus says that the four creatures of Revelation (Re 4:7) represent different aspects of the work of the Son of God, the Logos. They are key in connecting the work of the Logos between the Old and New Testaments, and it is clear that he also links this interpretation to Ezekiel's 'four-faced cherubim'[161]: The lion represents the imperial and royal aspects of the Logos when

---

[158] As noted in the section on the theological significance of the Chariot Vision below. 'Brightness' in other biblical passages denotes the glory of God. In Habakkuk 3:4 it is associated with the final judgement.

[159] Christman, A. R. (2005), 'What did Ezekiel See?,' Leiden: Brill, p. 19.

[160] Block, D. I. (1998), Ibid., p. 506.

[161] Christman, A. R. (2005), 'What did Ezekiel See?,' Leiden: Brill, p. 16.

He spoke to the patriarchs prior to Moses in the Old Testament and in the New Testament. He assigns the lion to the Gospel of John because of his glorious generation from the Father (Jn 1:1-14). The Calf represents the sacrificial and priestly role of the Logos. In the Old Testament the calf is the animal sacrificed by priests and in the New Testament it is seen in the Gospel of Luke, as it begins with the temple sacrifice of Zechariah (Lk 1:9). The human face represents Christ who in the Old Testament was foreshadowed by prophecy about Christ (for example the suffering servant of Is 42:1) and in the New Testament the Gospel of Matthew as he starts his gospel with the genealogy of Christ. Finally, the eagle represents the gift of the Spirit, in the Old Testament hovering over the surface of the deep at the Creation (Ge 1:2) and the gift of the Spirit in the New Testament. The Gospel of Mark is represented by the eagle since his first chapter manifests the spirit of the prophets. (Mk 1:2 quoting Is 40:3).[162]

Irenaeus says that there are four gospels, just as there are four regions of the world and four universal winds.[163] He says:[164]

> It is not possible that the Gospels can be either more or fewer in number than they are. For, since there are four zones of the world in which we live, and four principal winds, while the... 'pillar and ground' of the Church is the Gospel and the spirit of life; it is fitting that she should have four pillars, breathing out immortality on every side.... He who was manifested to men, has given us the Gospel under four aspects but bound together by one Spirit...

> For the cherubim, too, were four-faced, and their faces were images of the dispensation of the Son of God. The first living creature was like a lion, symbolizing His effectual working, His leadership, and royal power; the second was like a calf, signifying His sacrificial and sacerdotal order; but the third had, as it were, the face as of a man – an evident description of His advent as a human being; the fourth was like a flying eagle, pointing out the gift of the Spirit hovering with His wings over the Church. And therefore, the Gospels are in accord with these things, among which Christ Jesus is seated.

---

162 Irenaeus, *Adversus haereses*, III.11.8.

163 Irenaeus, *Adversus haereses*, III.11.7.

164 Irenaeus, *Ibid.*, 3.11.8.

This fourfold nature of proclamation of the Gospel reinforces the unity of the Gospel in Irenaeus' mind, and points to Christ.

Origen (184–253)

Origen sees the book of Ezekiel to be of and for the Church.[165] He sees Ezekiel as Christ. He says 'If you wish to hear Ezekiel, the 'son of man' preaching in captivity, understand him as a type of Christ'.[166] Origen sees the opening verse of Ezekiel: 'In the thirtieth year...' (Ez 1:1) as referring to Ezekiel's age. Modern commentators[167] agree with this. This links Ezekiel with Christ who began his public ministry at the age of 30. Origen says: 'in their thirtieth years the heavens were opened to both the Saviour and the prophet.'[168] Ezekiel is also called 'son of man' repeatedly (Ez 6:2; 12:2,8,18,27; 14:13; 16:2), and is the only prophet to be called in this way. Christman says that Origen's typology of seeing Christ as the key to Scripture continues to undergird all Christian exploration of Ez 1.[169]

Eusebius (263–339)

Eusebius identifies the likeness of the man seated on the throne (Ez 1: 26) as the Word. He says that the Glory of God (Ez 1:16-18) is a vision of the Word:

> Do you see how the passage (Ez 1:26) conceived of the glory of the Lord, the
> glory borne on the throne when it expounded the vision of the man? What

---

[165] Christman, A. R. (2005), *'What did Ezekiel See?,'* Leiden: Brill, p. 28 (footnote).

[166] Origen (2010), *Homilies 1–14 on Ezekiel,* trans Sheck, T. P., New York: Newman Press, p. 32.

[167] Block, D. I. (1997), *The Book of Ezekiel Chapters 1–24,* Grand Rapids: Eerdmans, p. 82.

[168] Origen (2010), *Ibid.,* p. 32.

[169] Christman, A. R. (2005), *Ibid.,* p. 29.

would this vision of a man be - which is said to be not God himself but the glory of God- what would it be except the only begotten Word of God."[170]

Eusebius sees in the 'likeness of a man' with *hasmal* above his loins, and fire below as being consistent with the Incarnation. The *hasmal* (which is equated with *electrum*, a naturally occurring alloy of gold and silver and was the most precious metal), is consistent with Christ's divinity.

Eusebius sees the mercy seat of the Ark of the covenant as 'a type and icon of the One seen above the cherubim in the prophet Ezekiel', whom he identifies with the Logos.*[171] The mercy (atonement) seat was made of gold and was overshadowed by the wings of two cherubim.

A wheel can refer to the whole of life for Eusebius. He therefore sees the wheel within a wheel (Ez 1:16) as the ever-changing life of human beings.[172] Similar to Athanasius and Cyril of Alexandria, Eusebius finds that the wheel within a wheel prefigures the spread of the gospel around the world.[173] Eusebius also sees the fire below the loins of the man seated on the throne (Ez 1:26) as reflecting the human sphere, particularly sexual activity.[174]

---

[170] Migne, J.-P. (1857–66), *Patrologiae Cursus Completus Series Graeca*, Paris: Imprimerie Catholique, 23.956b

[171] The Hebrew word for the cover of the ark or mercy seat, *hakappōret* (הַכַּפֹּרֶת) comes from the hebrew root *kāpar* (כָּפַר) which means both to cover over and to make propitiation. Eusebius connects the 'likeness of a man' with the Logos through this interpretation of the mercy seat by linking it with Paul's description of Christ in Ro 3:25: 'Christ Jesus whom God put forward as a sacrifice of atonement by his blood'. Cf.: Christman, A. R. (2005), *Ibid.*, p. 31.

[172] Migne, J.-P. (1857–66), *Ibid.*, 23.897c–d.

[173] Christman, A. R. (20.05) *'What did Ezekiel See?,'* Leiden: Brill, p. 36.

[174] Christman, A. R. (2005), *Ibid.*, p.30. (footnote).

Ambrose of Milan (337–397)

Ambrose initiated the theme of the wheel within a wheel being the New and Old Testaments... a wheel within a wheel is life under the law life under grace, inasmuch as Jews are within the Church and the Church within grace.[175]

St Jerome (347–420)

Jerome sees an analogy between Ezekiel being beside the river Chebar and the baptism of Jesus in the river Jordan.[176] Several authors, particularly Jerome, follow Irenaeus in ascribing the four Creatures as symbolising the Gospel writers. The interpretation of Jerome differs from that of Irenaeus for the lion and the eagle but is the same for the calf and for the man. Jerome chooses the lion for Mark because Mark's gospel begins with Isaiah's prophecy about John the Baptist, 'the voice of one crying in the wilderness, prepare the way of the Lord' (Jn 1:23). He chooses the eagle for John because of its lofty opening[177] and spiritual content.[178] Jerome's interpretation has been preferred to that of Irenaeus and is the standard interpretation of the four Creatures in Christian literature and art.[179]

Jerome sees the moving combination of the four Creatures and the wheels as showing the relationship of the gospels to each other and to the world. He quotes Ez 1:9 'and their wings touched one another' concerning the Living Creatures as showing that the gospels are, as he says 'joined together and cling

---

[175] Ambrose (1953), *De Spiritu sancto*, 3.21.162; in *Corpus Christianorum*, series Latina, Turnhout: Editions Brépois, 62, 81.13–28.

[176] St Jerome (2017), *Commentary on Ezekiel*, Ancient Christian Writers, New York: The Newman Press, p. 17.

[177] 'In the beginning was the Word and the Word was with God and the Word was God,' Jn 1:1.

[178] St Jerome (2017), *Ibid.*, p. 22.

[179] Christman, A. R. (2005), '*What did Ezekiel See?*,' Leiden: Brill, p. 18.

to one another and they run to and fro, soaring through the world';[180] 'they are always going to higher levels'.[181] In a commentary of Ez 1:12 'and each went straight forward' Jerome says that this forward straight movement reflects:

> the 'mystery of the Old and New Testaments because in those four animals, both the Law and the Gospel hasten on to what will be and never move backward.'

He says that what is true for the Gospels is also true for the virtues of the soul which always let go of what has gone before and hasten to what lies ahead'.[182]

Theodoret of Cyrus (393–457)

Theodoret held that the *hasmal* (electrum) was an analogy for Christ, with his two natures represented by the gold and silver of electrum.[183] The lower part of the figure with the 'likeness of a man' (Ez 1:26), Theodoret takes as Christ's divine nature whereas the upper half is his human nature. The fiery halo around the upper body in Ez 1:27: 'like the appearance of fire enclosed round about,' Theodoret takes to be a sign of the effect of the Incarnation on Christ's human nature.[184]

St Gregory the Great (540–604)

Gregory uses the interpretation of the four creatures given by Irenaeus (lion-Christ's imperial aspect, calf - sacrificial and priest, man - his incarnation,

---

[180] Migne, J.-P. (1841–64), Jerome, Epistles, 53.8, *Patrologiae Cursus completus,* Series Latina, Paris: Imprimérie Catholique.

[181] Jerome, (1953–), Commentary on Matthew, *Corpus Christianorum Series Latina,* Turnhout: Editions Brépois, 77, 3.76.

[182] Jerome (1953–), Commentary on Ezekiel, I.1.12. *Corpus Christianorum Series Latina,* Turnhout: Editions Brépois, 75, 16.360-17.372.

[183] Migne (1857-66), *Patrologiae cursus completus,* Series Graeca, Paris: Imprimérie Catholique, 81.836b.

[184] Zimmerli states that this phrase in Ez 1:27 was a later addition by the school of Ezekiel. (Zimmerli, W. (1979), *Ezekiel I,* Philadelphia: Fortress Press, p. 131.

eagle -spirit). Like Jerome, he takes the lion to stand for Mark (because of his opening quotation of Is 40:3); The calf stands for Luke (because his gospel begins with Zechariah's sacrifice); The man is for Matthew (because he starts with the genealogy of Jesus); The eagle is for John (because the opening of John focuses on the divinity of Christ). Gregory sees an eschatological meaning in the four creatures, because they emerge from a fiery cloud with brightness.*[185] Gregory supports this by several quotations from the Bible describing the End of Time, which have 'fire', surrounding storm, and 'brightness'.*[186] Gregory concludes that the 'cherubim' represent the saints who have been perfected through the gospels.[187]

Gregory also sees Christ in the four faces of the Creatures: he was a man, he died like a calf, rose like a lion, ascended like an eagle. In doing this he connects the New Testament with the Old Testament. He also sees a connection with the Church in that they are an analogy for the different ways in which the faithful imitate Christ and look forward to their fulfilment in him at the end of time. The perfected Christian becomes all four animals insofar as the person conforms to Christ in terms of the four qualities of the faces.[188] Gregory says 'all those who now have been perfected in the Church learned the righteousness of their perfection through the gospels'.[189] Gregory also sees, in the combination

---

[185] 'Brightness' in Ha 3:4 is associated with the final judgement.

[186] Christman, A. R. (2005), *'What did Ezekiel See?,'* Leiden: Brill, p. 18. The following passages are quoted by Gregory to support his eschatological interpretation: 1Co 3:13, 'for the Day will disclose it, because it will be revealed by fire, and the fire will test what sort of work each one has done'. 2Pe 3:10 'But the day of the Lord will come like a thief, and then the heavens will pass away with a loud noise, and the elements will be dissolved with fire..'(which both contain the word 'fire'). Ps 49:3 'Our God comes, he does not keep silence, before him is a devouring fire, round about him a mighty tempest' (which includes fire and a surrounding tempest), and Mt 24:27 'For as the lightning comes from the east and shines as far as the west, so will be the coming of the Son of man.' (uses the term *fulgur).*

[187] Christman, A. R. (2005), *Ibid.,* p. 19.

[188] Christman, A. R. (2005), *Ibid.* p. 23.

[189] Gregory the Great, *Homilies on Ezekiel,* I. 4. 2.

of silver and gold that go to form eleċtrum, the divine and human persons in Chriſt. Gregory sees the 'brightness around about' as the spread of the gospel.[190] He sees the wheel as symbolising the Bible. The wheel within a wheel (Ez 1:16) is the New Teſtament within the Old. Gregory says:

> A wheel was running within a wheel, and it was not hindered, the New Teſtament within the Old. It ran within that through which it was announced.

The wheels can also be a sign of the universal proclamation of the Gospel, and its growth in individual Chriſtians. Gregory sees Ez 1:20 'for the spirit was in the wheels' as signifying a spiritual co-operation between the Bible and the reader.[191]

Summary of Patristic Interpretations

The input of the Fathers into the interpretation of the Chariot Vision focuses particularly on the Living Creatures and the wheels. The unique input of the Fathers is that they see this ancient text as foreshadowing the New Teſtament. Almoſt all of them, ſtarting with Irenaeus, focus on the Living Creatures and see them as a link between the Old Teſtament and the four gospels. In doing this they are relating the four faces of the Creatures to a psychological anthropomorphising of these faces. The rationale that Jerome gives for the links to the four Evangeliſts has become so accepted by the Catholic Church that the Evangeliſts are rarely piċtured without their attributed Creature, as an insight to the charaċter of their writings. Gregory also sees the double wheels as representing the New Teſtament within the Old. Both Jerome and Gregory see an indication in the Chariot of spiritual progress over time.

Jerome is unique in seeing the Creatures as decision-making forces in man, and includes the eagle as representing conscience. He also sees the two levels of the 'figure like a man' as relating to different innate forces within man. Gregory

---

[190] Gregory the Great, *Ibid.* I.3.5
[191] Gregory the Great, *Homilies on Ezekiel*, I.6.12

sees an eschatological meaning in the Creatures. Jerome accepts that there are hidden meanings of the Vision that still need interpretation. Origen sees Ezekiel very much as a type of Christ and of the Church and sees the Chariot Vision in this light.

In conclusion therefore, the Fathers open a new chapter in interpretation of the Chariot Vision that shows that it is a very important link between the Old and New Testaments. They point the way to further exploration of the vision incorporating a psychological understanding of the Living Creatures and of what they mean for the Church.

Jewish Ancient and Medieval reception of the Chariot Vision in Ezekiel 1

Ancient reception of the Merkavah

One of the most ancient indications of the sacredness of Ezekiel 1 for the Jews was the synagogue practice of reading Ezekiel 1 with the Book of Exodus account of the theophany on Sinai and the giving of the Ten Commandments to Moses[192] on the feast of Shavuot.[193] Shavuot celebrates YHWH's revelation to the Jews and the acceptance by them of God's laws. The coupling of Ezekiel with this fundamental and sacred Torah reading shows the respect with which Ezekiel 1 was held by Jews. The perception is that these two texts belong together and must be read in the light of each other. Psalm 68 joins the concept of YHWH riding a chariot and coming from his holy place in Sinai. Halperin says that Ezekiel 1's vivid imagery 'gave people a starting point for visualising 'with the eyes of the heart', the full glory of the Sinai revelation'.[194] It also

---

[192]   Ex 19:1–20:23. See Appendix 14.

[193]   These practices were set up in the Talmudic era and have remain unchanged since the second century AD.

[194]   Halperin, D. J. (1988), *The Faces of the Chariot: Early Jewish Responses to Ezekiel's Vision*, Tubingen: Mohr Siebeck, p. 447.

incorporated an element of heavenly ascension that became the hallmark of those who meditated on the Merkavah.[195]

Medieval reception of the Chariot
Dangers of the Merkavah

In the Medieval interpretation of the Chariot ambiguities arose that caused certain rabbis to discourage those without careful teaching to meditate on the Merkabah. Two main concerns were that firstly, hidden within the Living Creatures (*hayyot*) were demonic, hostile forces that YHWH had brought under control, but could be a source of deception. For example, if the *hayyot* were actual intercessors between God and man, might not the ox have been the origin of the molten calf which seemed to have brought itself, out of the molten gold (Ex 32:24). Secondly the description of 'terrible ice' of the firmament in Ez 1:22 was the origin of endowing the crystal firmament with an inherent negative force, as an embodiment of chaos, also associated with the Red Sea. This again detracted from a true spiritual contemplation of the Merkabah. The perceived danger of the inadequately trained was that they might cause their own demise by angering YHWH, if they inadvertently introduced one of these alien distortions at a sacred point in their heavenly ascent. Meditation on the Merkavah and delving into its secrets was therefore proscribed for all except those who had gone through lengthy training.[196] Halperin feels that this attempt by rabbis to restrict the Merkabah generated excitement about this passage and gave 'its faith-affirming aspects all the more power.'[197]

---

[195]   Halperin, D. J. (1988) *Ibid.* p. 37.

[196]   Maimonides repeatedly reiterates this and states that when discussing the *merkabah* it was necessary to reveal some things and conceal others: Davies, D. (2011), *Method and Metaphysics in Maimonides' Guide for the Perplexed*, Oxford: Oxford University Press, p. 106.

[197]   Halperin, D. J. (1988), *Ibid.*, p. 450.

Philosophy versus Prophecy[198]

Maimonides' concept of divine science in the Chariot Vision appears to have come from Aristotelian metaphysics which was well developed in the Islamic world at the time.[199] Although Maimonides holds that Ezekiel's vision is a prophetic vision he gives a reading of the vision through the eyes of philosophy (divine science), seeing the *hayyot* as representatives of the heavenly spheres.[200] Although many Jews were hostile to philosophy, since philosophers rejected God's knowledge, Maimonides saw philosophy and religion as mutually complementary,[201] but he wished to distinguish knowledge gained through natural reason from knowledge gained by prophecy.[202] Maimonides says that readers of scripture who do not take it literally but make an effort to understand the meaning of scripture interpret scripture in the light of philosophical truth.[203] The Jewish medieval interpretation of the Merkabah therefore represented a tension between a prophetic interpretation as expounded in the Jewish tradition, and a 'philosophical' interpretation.

The Chariot Vision Current Interpretations

Block says that no other theophany in the entire Old Testament matches Ezekiel's inaugural vision. He says that the hope of Ezekiel's message for Christians is that in Jesus, the Messiah, the Glory of YHWH has descended, and

---

[198] Davies, D. (2011) *Ibid.*, p. 116.

[199] Kriesel, H. (2015), *Judaism as Philosophy: Studies in Maimonides and the Medieval Jewish Philosophers of Provence*, Boston: Academic Studies Press, p. 260.

[200] Kriesel, H. (2015), *Judaism as Philosophy: Studies in Maimonides and the Medieval Jewish Philosophers of Provence*, Boston: Academic Studies Press, p. 248.

[201] Davies, D. (2011), *Method and Metaphysics in Maimonides' Guide for the Perplexed*, Oxford: Oxford University Press, p. 157.

[202] Davies, D. (2011), *Ibid.*, p. 116.

[203] Davies, D. (2011), *Ibid.*, p. 187 (footnote 2).

dwells among us full of grace and truth (Jn 1:14). He is Immanuel.[204] Greenberg's interpretation of the Chariot Vision is that the first two appearances of the Chariot Vision (Ez 1:4 and 3:23) are private, whereas the appearances in Ez 8:2–11:23 and Ez 43:2 are visions that Ezekiel was given to communicate to others. He says that Ezekiel was initially dejected by the fate of his people and cast out by his community. Similar to the appearances in the Pentateuch of the divine majesty to Moses, when the people rise against him, (Ex 33:18) God appears to Ezekiel to vindicate his nonconformist position, and to give him support and show him favour. The overpowering vision allowed Ezekiel to brave the difficulties of his new role. Without the vision and the voice of God which followed (Ez 2:3), by which Ezekiel knew he had seen the Glory of God, he would not have had the courage to undertake his assigned role.[205] Boadt says that the Chariot Vision is of a divine throne occupied by God.[206] Eichrodt says that the Chariot Vision symbolises divine perfection, power and authority. The vision is a reproduction of the glory of the power of God.[207] Clements says that the prophet is convinced that God in person has appeared to him and this is what gives him the authority to speak to his fellow exiles.[208] Greenberg says that since the Vision has four wheels, this is more typical of a cart than a chariot, but primitive chariots also had four wheels. The chariot of the vision appears to be a combination of both cart and chariot.[209] He says that Ezekiel only appreciated that he was looking at the majesty of God when

---

[204] Block, D. I. (1998), *The Book of Ezekiel Chapters 24–48*, Grand Rapids: Eerdmans, p. 746.

[205] Greenberg, M. (1983), *Ezekiel 1–20*, New York: Doubleday, p. 81.

[206] Boadt, L. (1990), 'Ezekiel,' *New Jerome Biblical Commentary*, London: Burns and Oates, p. 310.

[207] Eichrodt, W. (1970), *Ezekiel*, London: SCM Press, p. 58.

[208] Clements, R. E. (1996), *Ezekiel*, Westminster Bible Companion, Louisville: Westminster John Knox Press, p. 11.

[209] Greenberg, M. (1983), *Ibid.*, p. 57.

he saw the fiery human figure on the top (Ez 1:25-28).[210] There are few analogies in scripture,*[211] the closest being Ps 18:8–14:

> Then the earth reeled and rocked; The foundations also of the mountains trembled and quaked, because he was angry. Smoke went up from his nostrils and devouring fire from his mouth; Glowing coals flamed forth from him. He bowed the heavens and came down; thick darkness was under his feet. He rode on a cherub, and flew; he came swiftly on the wings of the wind. He made darkness his covering around him, his canopy thick clouds dark with water. Out of the darkness before him there broke through his clouds hailstones and coals of fire. The Lord also thundered in the heavens, And the Most High uttered his voice, hailstones and coals of fire. And he sent out his arrows and scattered them; He flashed forth lightnings and routed them.

He says that the meaning of the vision for the prophet may have been to 'revolutionise' a notion that YHWH's revelation could not occur outside of the land of Israel.[212] Brownlee says that the Chariot Vision meaning is that the cosmic Lord of the universe is intervening in history to judge Israel and to warn them through one man, Ezekiel. The Vision is only an image of the infinite.[213] Joyce says that the Chariot Vision is nothing less than a vision of God himself. The first chapter of Ezekiel has been given great importance in Christianity, but particularly in Judaism.[214] Odell says that the vision should be understood as 'a highly stylized mediation on God's commanding presence in the world'.[215] The Living Creatures represent the powers of the created order, and also the possibility that such power can rebel against God, but they are

---

[210] Greenberg, M. (1983), *Ibid.*, p. 52.

[211] But see also Appendix 1: The Book of Enoch 14. Enoch's heavenly ascent.

[212] Greenberg, M. (1983), *Ibid.*, p. 59.

[213] Brownlee, W. H. (1986), *Ezekiel 1–19*, World Biblical Commentary, Volume 28, Waco: Word, p. 18.

[214] Joyce, P. M. (2007), *Ezekiel: A Commentary*, New York, T & T Clark, p. 65.

[215] Odell, S. (2005), *Ezekiel*, Smyth & Helwys Bible Commentary, Macon: Smyth & Helwys, p. 32.

now subordinated to the enthroned glory and ready to do its will. The Vision also emphasizes God's relationship with the whole cosmos. It does not present God in abstraction from the cosmos but in relation to it. Odell also states that Ezekiel's vision has implications for theological method; she sees the logic of this theology is 'to take the metaphors of earthly domination and political hegemony and to transpose them into a transcendental vision of divine rule that relativizes all other claims to power' – since all other powers, like the living beings must eventually submit to God's rule. Ezekiel's vision turns our attention from chaos to calm.[216] Odell says that the Vision is a representation, of the form of the divine glory and not the divine glory itself.

Zimmerli says the vision shows 'the continuity of divine faithfulness' and proclaims the absolute sovereignty of the God of Israel even in a foreign land. This manifestation of divine faithfulness and sovereignty are also announcements of God's grace. The prophet encountered God as the Lord who had already earlier revealed himself to his people Israel, in storm and light (Ex 19:1).[217] 'Ezekiel was called to be a witness to a history which the Christian Church believes has its centre in Jesus Christ.'[218]

Two further key appearances of the Chariot Vision are reviewed by Block. In Ez 11:22 the Glory of YHWH ascends from the middle of the city and stops on the mountain east of the city, the Mount of Olives. It is unclear where it goes, but Block says that from the Temple, the Mount of Olives represents the eastern horizon. Later tradition seems to assume that the throne-chariot takes 'the Glory' back to its real and eternal home in heaven. The vision of the Ancient of Days (Da 7:9-10) sitting on a wheeled throne seems to support this.[219] This departure marks the end of the theophany, although twenty years later (when Ezekiel is 50 years old) the glory of YHWH returns into the Idealised Temple

---

[216] Odell, S. (2005), *Ibid.*, p. 37.

[217] Zimmerli, W. (1979), *Ezekiel 1*, Philadelphia: Fortress Press, p. 141.

[218] Zimmerli, W. (1979), *Ibid.*, p. 142.

[219] Block, D. I. (1997), *The Book of Ezekiel Chapters 1-24*, Grand Rapids: Eerdmans, p. 359.

Vision from the east and enters through the East Gate. The earth 'shone with his glory' (Ez 43:2) and his glory filled the temple: compared with the initial vision, the radiance of the divine glory seems more intense.[220] The prophet falls on his face as at the first appearance of the vision. The Living Creatures are not mentioned in this vision.

The divine speech given to Ezekiel on arrival of the glory of YHWH into the temple is an important element in understanding the Chariot Vision. Block divides it into two parts: an explanation and a charge to the prophet. The explanation is that YHWH declares that the temple is his palace and asserts kingship over Israel. Ezekiel portrays that it is the temple rather than the ark of the covenant that is now the throne of YHWH: 'the place for the soles of my feet' (Ez 43:7). Block says that this fulfils Jeremiah's prediction: 'in those days... they shall no longer say, 'the ark of the covenant of the Lord'. At that time Jerusalem shall be called the throne of the Lord, and all nations shall gather to it...' (Je 3:17). The existence of the city will be evidence of YHWH's eternal presence (Ez 48:35).

YHWH demands a holy reputation: He will not tolerate an unholy people misrepresenting him before the nations. Through Ezekiel, the house of Israel is charged to end their idolatry and four specific behaviours that have defiled the holy name of YHWH: Firstly, the sins of the people (their harlotry);[221] secondly, cultic abominations especially in relation to kings; thirdly, violation of the sacred space of the temple with profane constructions; fourthly, general pagan activities. The speech ends with an exhortation to guard the sanctity of the people and the temple and the reputation of God. YHWH will then return and establish permanent residence in the midst of his people.[222] He

---

[220] Block, D. I. (1998), *The Book of Ezekiel Chapters 24–48,* Grand Rapids: Eerdmans, p. 579. The intensity of this vision might reflect the fact that this was the first (and only) appearance of the Glory of YHWH after the Fall of Jerusalem and destruction of the Temple and occurred three years before Ezekiel's death.

[221] Block, D. I.(1998), *Ibid.,* p. 582.

[222] Block, D. I.(1998), *Ibid.,* p. 586.

then declares the utter sanctity of the temple in giving a new Torah of the temple (Ez 43:12- 46:24). The new Torah sets rules for atonement (Ez 43:20) that reflect the Levitical rules of atonement (Le 16).

The vision of the return of the Glory of God to the temple proclaims three things:[223] 1. The ineffable holiness of God; 2. The glorious mercy of God; 3. The sinfulness of human beings. Block says that the relevance of the new Torah for the modern believer is that true worshippers fall down before the transcendent majesty of God and proclaim his holiness and that they joyfully accept his gracious invitation to eat the communion meal, the Eucharist, in his presence.[224]

Conclusion of Chapter

The arguments given here show that the Chariot Vision should be treated as objective privileged revelation to Ezekiel that redounds upon and informs the whole book. To Ezekiel, the Chariot Vision enfolded the glory and actual presence of YHWH. The Patristic interpretations bring out two major strands that are central to this thesis: (a) the very early understanding of the Living Creatures as anthropomorphised character forces and (b) the deeply embedded Christological focus of Ezekiel and the vision. For Maimonides, divine science was the application of reason to understanding scripture and complemented prophetic revelation and was not in opposition to it. Modern commentators see the vision as a symbol or image with a message for his people. The successive appearances and departures of the vision give a subtext that links the two ends of the book with a holiness imperative over time as well as giving an eschatological perspective.

---

[223]   Block, D. I. (1998), *Ibid.*, p. 590.

[224]   Block, D. I.(1998), *The Book of Ezekiel Chapters 24–48*, Grand Rapids: Eerdmans, p. 686.

Chapter 3

Middle Tier: The Living Creatures

Overview of Chapter

This chapter first looks at the background of the Living Creatures, what they are and represent, their moral significance and their Hebrew descriptors. Secondly it reviews their potential Mesopotamian origins within the development of Babylonian constellations and astronomy/astrology and the origins of the zodiac. Thirdly it explores the background to the choice of the faces and their assigned characteristic traits as determined over time. Fourthly it explores how their fixed relative positions in the Chariot Vision can be a metaphor for decision making. A parallel is sought between the interaction of the Creatures' character forces and modern psychological decision making theory. The significance of the Living Creatures straight movements is sought as an interaction between spirit and reason.

The Middle tier of the Chariot Vision is clearly delimited above by the crystal firmament (Ez 1:22). There is not a clear separation of the middle tier from lower tier, but the wheels are described as being 'below the cherubim': *'ĕl-tāḥat* (אֶל־תַּחַת) Ez 10:2. The wheels are therefore in the lower tier. The wheels are also described as being beside the Living Creatures (as in Ez 1:15) but for purposes of this thesis they constitute the lower tier of the Chariot Vision. Fire and coals appear to extend both between the Living Creatures and between the wheels, and so appear to occupy both lower 'tiers'.

The composite beings with four faces, four wings a human body and hands and calves' feet are probably the most enigmatic part of the Chariot Vision. Principal questions that arise about them are: 1) do they represent the spiritual world, temporal world, or both? 2) What does Ezekiel's use of composite creatures signify and does it reflect anything in the *Sitz im Leben* of Ezekiel? 3) Is there a specific figurative meaning to the four different faces?

After discussing these questions, a short review of authors who have attributed a moral function to the Living Creatures is presented. Other elements in the middle tier will also be reviewed: the bodies and wings of the Creatures and the 'lightning' that flashes between the Creatures.

Do the Creatures Represent Spiritual or Temporal Beings?

At their first appearance of the Chariot (Ez 1:5) the word *ḥayyôt* which is translated as 'Living Creatures' is used for the composite beings. However, at their second appearance, the word *kĕrûbîm* is used. Block says that Ezekiel may not have recognised the Living Creatures as cherubim at their first appearance, but when he sees the Chariot Vision again in Ez 10:1, he is able to recognise them as cherubim, a type of angel.[225] Halperin however, feels that calling the Living Creatures 'cherub' is a later addition that was put in to justify a tendency in 𝕸 to make the wheels a class of angels.[226] It is discussed below, that since the Creatures are able to function together, they have human reason. According to Aquinas,[227] angels do not have reason. In this thesis therefore, they are not treated as a type of angel, and are called 'Living Creatures'.

---

[225] Block, D. I. (1997), *The Book of Ezekiel chapters 1–24*, Grand Rapids: Eerdmans, p. 320.

[226] Halperin, D. J. (1976), 'The Exegetical Character of Ezekiel x 9–17,' *VT*, **26**, 129–41; p. 137.

[227] Aquinas, *Summa Theologiae*, II 2 49.

The fact that the Living Creatures have wings as well as human and animal characteristics, symbolizes that they moved between the divine and earthly realms. By analogical thinking, the four Creatures could only function at a level higher than earthly reality.[228] Eichrodt says that Ezekiel stresses the human resemblance of the Living Creatures (with the man's face in front and the form of a human body and hands),[229] and that they go between the divine realm and the human realm. This suggests that they may play a role in revelation.

Background to composite creatures

Most commentators see the Living Creatures as an analogy, that resonates with Mesopotamian art. Gods that had both animal and divine features were common in Egypt and Mesopotamia.[230] Block says that in the four Creatures, YHWH is adopting the art of Mesopotamia that was contemporary to Ezekiel. He says that while extra-Israelite motifs have been incorporated into the Chariot Vision, this does not represent capitulation to pagan thought, but in fact challenges pagan conceptions.[231] Whereas YHWH had chosen the language of Canaan with which to communicate verbally (Is 19:18), he now uses the art of Mesopotamia to communicate visually. No analogues to creatures with four faces have been found although Allen illustrates an eighteenth-century Assyrian god representation with four human faces.[232] However, multiplication of heads is extremely rare in Mesopotamian art.[233] Sweeney says that Ancient Near Eastern art portrayed cherubim as guardians of royal thrones, temples

---

228 Launderville, D. (2004), 'Ezekiel's Throne-Chariot Vision: Spiritualizing the Model of Divine Royal Rule,' *Catholic Biblical Quarterly*, **66**, 361–377, p. 374. Analogical thinking is based on the idea that because two or more things are similar in some respects, they are probably also similar in some further respect.

229 Eichrodt, W. (1970), *Ezekiel*, London: SCM Press, p. 55.

230 Greenberg, M. (1983) *Ezekiel 1–20*, New York: Doubleday, p. 55.

231 Block, D. I.(1997), *The Book of Ezekiel Chapters 1–24*, Grand Rapids: Eerdmans, p.108.

232 Allen, L. C. (1994), *Ezekiel 1–19*, World Biblical Commentary 28, Dallas: Word, p. 29.

233 Greenberg, M. (1983) *Ibid.*, p. 55.

and city gates. Boadt says that much of the imagery is borrowed from typical Babylonian artistic descriptions of divine beings, but also sees connections to the ark of the covenant and its protective cherubim found in the Holy of Holies. The use of 'living' *ḥay* (חַי) suggests quasi-divine beings which are well known in Assyrian art as guardian deities of gates and palaces and of the king's person.[234] Odell suggests that the Living Creatures could be an amalgam of four Assyrian demon figures: *umu-apkallu* (or a winged human); *kusarikku* (or a bull-man with legs of a bull and a human face and trunk); *ugallu* (or a lion-demon with a human body and a lion's head) and *apkallu* (with a human body and the head of a bird).[235]

Choice of faces of animals

Greenberg says that the faces of the Living Creatures are used to express their attributes.[236] The faces of the Living Creatures in the initial Chariot Vision were of a man, a lion, an ox and an eagle (Ez 1:10). Jerome cites Hippocrates as saying that the four faces link to the four elements : 'fire, air, water and earth, out of which everything consist'.[237] Jewish biblical commentaries emphasize that the animal imagery is not to be taken literally but provides analogies for the various ways that God 'reveals himself to the world'.[238] Block says that the selection of these particular animals, was natural for the world in which Ezekiel lived because these animals appeared frequently on ancient iconographic and glyptic art and had symbolic significance for the Israelites: The lion served as a

[234] Boadt, L. (1990), 'Ezekiel,' *New Jerome Biblical Commentary*, London: Burns & Oates, p. 310.

[235] Odell, S. (2005), *Ezekiel*, Smyth & Helwys Bible Commentary, Macon: Smyth & Helwys, p. 27.

[236] Greenberg, M. (1983), *Ezekiel 1-20*, New York: Doubleday, p. 55

[237] St Jerome (2017), *Commentary on Ezekiel, Ancient Christian Writers*, New York: The Newman Press, p.23.

[238] 'Merkavah, A Hasidic Explanation,' *New World Encyclopedia*, 2016, at *http://www.new-worldencyclopedia.org/entry/merkavah#A_Hasidic_explanation* Accessed 31/12/19.

symbol of royalty, being strong, fierce and courageous. The ox was a symbol of fertility and divinity, and the eagle was the swiftest and most stately of birds. The human was created in the image of God and was invested with divine majesty and was the most noble and dignified of all.[239] Block says that the significance of these creatures exceeds their sum since they carry the divine throne and in doing so declare that YHWH has the strength and majesty of the lion,[240] the swiftness and mobility of the eagle, the procreative power of the bull and the wisdom and reason of mankind.[241] Blenkinsopp says that the four faces of the Living Creatures are typical representatives of all of creation and draw on motifs that are familiar in Near East iconography.[242] Allen explores the figurative use of each of the four animals who make up the Living Creatures. The lion is the fiercest of beasts; the eagle is the most imposing of birds; the bull is the most valued of domestic animals (for ploughing and breeding); men rule them all.[243] Sweeney says that the four faces of the cherubim are: the human being which conveys divine intelligence and the knowledge of good and evil; the lion represents divine sovereignty, the royal tribe of Judah and the house of David; the bull divine strength and the eagle divine mobility throughout the world.[244] In Mesopotamia, there was a precedent of attributing desirable animal characteristics such as these to the king through the constellations.[245] This will be discussed further below.

---

[239] De 28:49 'as swift as an eagle'.

[240] Cf. Is 31:4 Like as the lion...so will the LORD of hosts come down to fight upon mount Zion,

[241] Block, D. I.(1997), *The Book of Ezekiel Chapters 1–24*, Grand Rapids: Eerdmans, p. 96.

[242] Blenkinsopp, J. (1990), 'Ezekiel,' *Interpretation: A Bible Commentary for Teaching and Preaching*, Louisville: John Knox Press, p. 21.

[243] Greenberg, M. (1983), *Ezekiel 1–20*, New York: Doubleday, p. 56.

[244] Sweeney, M. A. (2013), *Reading Ezekiel*, Macon: Smyth & Helwys, p. 28.

[245] White, G. (2007), *Babylonian Star-Lore*, London: Solaria, p. 184.

The Living Creatures and morality

The interpretation of the four Creatures as a moral comparator is very ancient. Jerome cites Plato as viewing the man, lion and calf as the rational, irascible and appetitive parts of the soul.*[246] Jerome himself says[247]

> Therefore, God guides this four-horse team like a charioteer, and he holds in check the one that runs with steps that are out of control, making it docile and compelling it to obey his command.

Several modern commentators have seen forces of morality within the Living Creatures. Launderville says that the being of each of the four creatures has to function within a division of power among the four faces of its head. He says they therefore need the spirit for them to be able to do anything, and by the same token they cannot rebel.[248] The combined characteristics of the Creatures gives them powers to be defenders of the divine throne.[249] Carley sees the cherubim (which means intercessors) as guardians of God's throne.[250] Odell says the fact that the living beings are directed by the spirit shows that

---

[246] St Jerome (2017), *Commentary on Ezekiel, Ancient Christian Writers*, New York: The Newman Press, p. 22: 'In accordance with Plato who calls them λογικόν, θυμικόν, ἐριθυμητικόν. (To the man) they assign reason, reflection, mind and counsel the same virtue and wisdom and locate it in the citadel of the brain; but they refer to the lion fierceness, anger and violence, things that are located in the gallbladder; furthermore, lust, excess and the desire for all pleasures is in the liver, that is in the calf, an animal that sticks to the works of the earth; they assign as the fourth principle that is above and beyond these here what the Greeks call συνείδησιν. This is the spark of conscience... By means of it we perceive that we are sinning when we are overcome by pleasures or by rage, sometimes when we are misled by something that closely resembles reason. They assign this particularly to the eagle...' Cf: Kries, D. (2002), 'Origen, Plato and Conscience (synderesis) in Jerome's Ezekiel Commentary,' *Traditio*, **57**, pp. 67–83; p. 68.

[247] St Jerome,(2017), *Ibid.*, p. 23.

[248] Launderville, D. (2004), 'Ezekiel's Throne-Chariot Vision: Spiritualizing the Model of Divine Royal Rule,' *Catholic Biblical Quarterly*, **66**, pp. 361–377; p. 366.

[249] Launderville, D. (2004), *Ibid.*, p. 365.

[250] Carley, K.W. (1974), *The Book of the Prophet Ezekiel*, Cambridge: Cambridge University Press, p. 61.

the living beings are part of the created order and that they move in harmony because God keeps all in perfect control and they have become servants to the enthroned glory.[251] She sees the living beings as four demons, three of which were associated with the powers and who attempted revolt against divine rule of the gods. The fourth was a bull-man, who was an adversary of the gods. By Ezekiel's time these demons were understood as spirits who had come under control and were ready to enact the wills of the gods. Each face therefore represented a different element of mastery of the forces of chaos.[252] The stiff-leggedness of the creatures was interpreted by Mackay[253] as suggesting a tendency to pride and opposition to divine rule. Based on a *midrash* Greenberg points out the four faces could represent different aspects of pride:[254]

> four kinds of proud beings were created in the world: the proudest of all-man;
> of birds- the eagle; of domestic animals – the ox and of wild animals- the lion.
> And all of them are stationed beneath the chariot of the Holy One...

Bodies and wings of the Living Creatures

The four creatures had an erect posture and had hands and a leg like humans. Although the text appears to give each creature only one leg *regel* (רֶגֶל), the full phrase here is 'and their legs, a straight leg' *wĕraglêhem regel yĕšārâ* (וְרַגְלֵיהֶם, רֶגֶל יְשָׁרָה). However, *regel yĕšārâ* although singular, could be meant in a collective sense. This phrase then becomes: 'And their legs, were "Straight-legged".' Mackay says their straight legs signifies they characteristically stood

---

[251] Odell, S. (2005), *Ezekiel*. Smyth & Helwys Bible Commentary. Macon: Smyth & Helwys, p. 32.

[252] Odell, S. (2005), *Ezekiel*, Smyth & Helwys Bible Commentary, Macon: Smyth and Helwys, p. 27.

[253] Mackay, J. L. (2018), *Ezekiel: A Mentor Commentary*, Vol. 1, Chapters 1–24, Fearn: Christian Focus Publications, p. 88.

[254] Greenberg, M. (1983), *Ezekiel 1–20*, New York: Doubleday, p. 56. (*midrash* to Ex 15:1).

upright and did not kneel in the divine presence.[255] Not kneeling in the divine presence could be because the posture is required by the Almighty. The Hebrew priests offering sacrifice did not kneel as they had to be occupied with making sacrifice at the altar. This suggests that the Living Creatures could be an analogy for priests. The Living Creatures go between divine and human realms, just as priests are intercessors between God and man. Gregory the Great had suggested that the traits of the four creatures are essential for the perfected Christian.[256] The Chariot Vision therefore may contain a sign that Ezekiel is initiating his priesthood.

The fact that the Creatures had a calf's foot meant that, unlike human feet which point in a particular direction, the cherubim had to be guided towards any direction.[257] They had a human form under their wings (Ez 10:8). Of their four wings, two were stretched out straight, one towards the other (Ez 1:23), the other two covered their 'bodies', *gĕvyōtêhem* (גְּוִיֹתֵיהֶם). The two outstretched wings of the angels in Isaiah were used for flight (Is 6:3). In Ezekiel the wings made a sound when the Creatures moved, but there is no clear statement that they flew (Ez 1:24). When they stopped, they let down their wings. The position of the outstretched wings touching each other is similar to those of the cherubim in the inner sanctuary of the temple of Solomon (1Ki 6:27). Zimmerli says that the position of the wings is therefore a ceremonial attitude.

The Sounds of the wings of the Living Creatures

In Ez 1:24, the sounds of the wings of the Living Creatures are: like the sound of 'mighty waters', *mayim rabbîm* (מַיִם רַבִּים), 'like the thunder of the Shaddai'

---

255   Mackay, J. L. (2018), *Ibid.*, p. 88.

256   The Living Creatures go between divine and human realms, just as priests are intercessors between God and man. Gregory the Great (*Homilies on Ezekiel*, I.4.2.) had suggested that the traits of the four creatures are essential for the perfected Christian.

257   Mackay, J.L. (2018) *Ezekiel: A Mentor Commentary*, Volume 1, Chapters 1–24, Fearn: Christian Focus Publications, p. 88.

kĕqôl šadday (כְּקוֹל־שַׁדַּי), a tumult*[258] 'like the sound of an army' *qôl hamullah kĕqôl maḥāneh* (קוֹל הֲמֻלָּה כְּקוֹל מַחֲנֶה).[259] In Ez 3:13 the sound of the wings of the Living Creatures and the wheels sounded 'like a great earthquake' *wĕqôl ra'aš gādôl* (וְקוֹל. רַעַשׁ גָּדוֹל). Although Ezekiel does not apply sounds to the wheels in Ez 1.15-1.21, patristic authors, particularly Eusebius of Caesarea, link this passage to Ps 77:19 'the sound of your thunder was in the whirlwind'.[260,*261] Cyril of Alexandria says that the thunder is the Gospel preaching resounding throughout the world, and links it to the wheels in Ez 1:16.[262]

The 'Lightning' that Flashes from the Fire between the Living Creatures

In the Bible, lightning is often a manifestation of a theophany: Lightning flashed in Ex 19:16 when Moses went up mount Sinai, and again after Moses received the ten commandments in Ex 20:18. In Re 11:19, after God's temple was opened, lightning flashed. Lightning is also used as a weapon against God's foes: 2Sa 22:15 'he sent forth arrows, and scattered them, and he flashed lightning, and dismayed them'; Ps 144:6 'flash your lightnings and scatter them'.

In Ezekiel the use of lightning appears to be neither of these. Although the Chariot Vision is effectively a theophany, the two places in which lightning are

---

[258] Strong's concordance gives the meaning of *hamullah* 'a rushing or roaring sound'.

[259] Zimmerli, W. (1979), *Ezekiel 1*, Philadelphia: Fortress Press. p. 130. Zimmerli says that the comparison of the sound of the wings of the cherubim in Ez 1:24 to the voice of Shaddai is a late addition since it is lacking in 𝕲. Also the comparison with the sound of a military camp is an addition. He says that the noise of the sea and the mysterious sound of water, with its supernatural associations is what drove the later redactors to make these comparisons

[260] Christman, A. R. (2005), *'What did Ezekiel See?,'* Leiden: Brill, p.34.

[261] It would be more logical however to link Ps 77 to Ez 10 where *galgal* is used to describe the wheels or to Ez 3:13 where the wheels make the sound of an earthquake.

[262] Migne, J.-P. (1857-66), *Patrologiae cursus completus series Graeca,* Paris: Impimérie Catholique, 69.1193.1-b.

mentioned relate to the fire between the Living Creatures: in Ez 1:13 'and out of the fire went forth lightning' *bārāq* (בָּרָק). The lightning comes from the fire between the Living Creatures. In the subsequent verse, Ez 1:14 it is used as a simile for the speed of movement of the Living Creatures: 'the Living Creatures darted back and forth like a flash of lightning *habbāzāq* (הַבָּזָק)'. These are two different but linked uses for 'lightning'. The two Hebrew words *bārāq* and *bāzāq* have very similar meanings: *bārāq* is lightning (*in general*). *Habbāzāq* is a hapax in Biblical Hebrew. In rabbinic writings it denotes 'to flash' (in the *niphal* form).[263] It therefore means a *bolt or flash* of lightning. The second dramatizes the movements of the Chariot. The first says something about the source of the lightning. Nowhere else is lightning described as coming from fire (except see subsequent paragraph below concerning Ez 1:4), and so it tells us something about the fire. In the chapter 'Spirit in Ezekiel and the Chariot,' that the fire is caused by, or is a manifestation of, the spirit is discussed. Similarly, the lightning is likely a manifestation of the spirit, since it originates from the fire, and is synchronous with the movement of the Living Creatures and wheels: 'and out of the fire went forth lightning. And the Living Creatures darted back and forth like a flash of lightning' (Ez 1:13-14).

In Ez 1:4 the Hebrew words *wě'ēš mitlaqqaḥat* (וְאֵשׁ מִתְלַקַּחַת) are also interpreted by Block as signifying lightning bolts.[264] The word *mitlaqqaḥat* in Ez 1:4 is a participle of the *hithpael* or reflexive form of the verb: *lāqaḥ* (לְקַח). The verb *lāqaḥ* in its first or *qal* form is translated variously as to take, to fetch or to lead. The reflexive form would therefore denote: to take, fetch or lead *oneself*. This implies a movement and is not a description. A participle can act as an adjective or a verb in Hebrew.[265] Since the only verb in this verse*[266]

---

263   Block, D. I. (1997), *The Book of Ezekiel Chapters 1–24*, Grand Rapids: Eerdmans, p. 95 (footnote 47).

264   Block, D. I. (1997), *Ibid.*, p. 93.

265   Kelley, P. H. (1992), *Biblical Hebrew*, Grand Rapids: Eerdmans, p. 200.

266   'Suddenly I *noticed* a stormy wind blowing in from the north – an enormous cloud with fire flashing back and forth and a radiance surrounding it.'

is at the beginning ('noticed') and refers to the object: wind, the participle *mitlaqqahat* is likely to be a verb with 'fire' as its subject. Therefore a likely interpretation would be fire 'moving itself'. This could be interpreted as lightning as Block suggests, but since there are two other words in Hebrew for lightning (*bārāq* and *bāzāq*) and both are used just later in Ez 1:13, 14 (see above) to describe something coming from the fire, and neither is used to describe the fire itself, it appears that Ezekiel was trying to express something different to lightning. This could be a fire moving by itself, changing its location, possibly zig-zagging. A repetitive movement of the fire is suggested by Ez 1:13, where another *hitphael* reflexive participle: *mithallēkĕt* (מִתְהַלֶּכֶת) from the *qal* form of *hālak* 'to go or come' (הָלַךְ) is used to describe the fire. *Mithallēkĕt* means 'to go back and forth'. A repetitive movement is also suggested in the same verse, by the use of *ratsô' washôv* (רָצוֹא וָשׁוֹב) meaning 'went and returned', to describe the Living Creatures' movements.*[267] Driver[268] says that *mitlaqqahat* likely means 'kindled' since the Syriac for 'kindled' *'ûbed* also has the reflexive sense of 'caused to catch on'. The lack of a sense of subjective movement in 'kindled' however is against this interpretation. The only other place *mitlaqqahat* is used in the bible is in Ex 9:24 when the Lord through Moses, sent thunder, hail and fire to rain down on the earth in the seventh plague. It is used to describe 'fire which flashed in the midst of the hail' *wĕ'ēš mitlaqqahat betôk habbārād* (וְאֵשׁ מִתְלַקַּחַת בְּתוֹךְ הַבָּרָד)

The question arises whether 'lightning' described in Ez 1:13 and 14 really is lightning, or a description of flashes from the fire that just looked like lightning? When Ezekiel first sees the fire flashing from a storm cloud, (Ez 1:4) the cloud is obviously at a distance. Brownlee says that the storm cloud is unusual in its

---

267 Talmud Mas. Chagigah 13a 3–5: 'And the Living Creatures ran and returned as the appearance of a flash of lightning: 3. What is the meaning of 'ran and returned'? — Rabbi Judah said: Like the flame that goes forth from the mouth of a furnace: 4. What is the meaning of 'as the appearance of a flash of lightning'? — Rabbi Hanina said: Like the flame that goes forth from between the potsherds': 5.

268 Driver, G. R. (1961), 'Ezekiel's Inaugural Vision,' *VT*, p. 60–62.

direction (in Mesopotamia storms come from the southwest not the north) and time of year (they usually occur in winter or spring not in July, which is the month calculated from the dates given in Ez 1:1-2).[269] When Ezekiel starts to describe details like the faces of the Creatures (Ez 1:6), the fire is obviously very much closer, and by Ez 1:15 a wheel is described as being on the earth. So the 'lightning' described in Ez 1:13 *wĕmîn ha'ēš yoṣe bārāq* (וּמִן־הָאֵשׁ יוֹצֵא בָרָק) and in Ez 1:14 *kĕmar'ēh habbāzāq* (כְּמַרְאֵה הַבָּזָק) is not likely real lightning, but just has its flashing appearance. What could the 'lightning' described in Ez 1:13 be physically? An electrostatic discharge travels a short distance and would give the a lightning-like visual effect at close range and could represent a visible transfer of the spirit to the Living Creatures. The 'fire moving itself' (*'ēš mitlaqqaḥat*) as well as the 'lightning' (*bārāq* and *bāzāq*) might therefore be manifestations of the spirit.

Mesopotamian Influences in the Chariot Vision

The Living Creatures and Divine Power

The probable date of Ezekiel's first vision is 31 July 593 BC.[270] Zedekiah had journeyed to meet with Nebuchadnezzar in that year (Je 51:59). The river Chebar was a man-made canal in Nippur, just South of Babylon. Nippur was an ancient city that was once on the Euphrates river. The Euphrates shifted its course and this led to a decline in Nippur's fortunes. Nippur had been under Assyrian control and had suffered a decline in population probably due to warfare and the Jews were probably exiled there in the sixth century, in order to supplement the population. Ezekiel had been living as an exile for

---

[269] Brownlee, W. H. (1986), *Ezekiel 1–19*, World Biblical Commentary Volume 28, Waco: Word, p. 10.

[270] Block, D. I. (1997), *The Book of Ezekiel Chapters 1–24*, Grand Rapids: Eerdmans, p. 26.

five years when he saw the first Chariot Vision. This is the background for the resemblance between the Book of Ezekiel and Assyrian traditions.[271]

The Living Creatures of the Chariot Vision show Mesopotamian influence. A mobile throne on wheels appears to derive from a widespread ancient oriental custom.[272] The portrayal of power by using a composite creature incorporating the strengths of different creatures was well established, the most widely known being the Lamassu (see subsection 'The Lamassu'). In Ezekiel's chariot the composite Living Creatures appear to serve to establish the political power of the 'Glory of God' in the *Sitz im Leben* of sixth century Babylonia.[273] Composite figures of animal and human were used for their apotropaic power to guard temples and houses starting in the Akkad period (2,350 BC).[274] The Living Creatures of the Chariot Vision are not an exact parallel of any known iconography in the Mesopotamian tradition, but anthropomorphic imagery was in use to denote political power in the Assyrian civilisation (1,350-610 BC) at the time of Ashurnasirpal II in Nimrud.*[275]

There are analogies in Mesopotamian art to a king being borne on creatures who are composite animal and human,[276] since Assyrian royalty had used

---

[271] Odell, S. (2005), *Ezekiel*, Smyth & Helwys Bible Commentary, Macon: Smyth & Helwys, p. 14.

[272] Dürr, L. (1917), *Ezekiels Vision von der Erscheinung Gottes (Ez.c. 1 und 10) im Lichte der vorderasiatischen Altertumskunde*, Wurtzburg: Richter, pp. 31–21.

[273] Launderville, D. (2004), 'Ezekiel's Throne-Chariot Vision: Spiritualizing the Model of Divine Royal Rule,' *Catholic Biblical Quarterly*, **66**, pp. 361–377; p. 376.

[274] Mischwesen, A. (1997), 'Philologisch. Mesopotamien,' in Wiggermann, F. A. M., *Reallexikon der Assyriologie (RlA)*, **8**, pp. 222–224; p. 231.

[275] Matthews, I. G. (1939), *Ezekiel*, American Commentary on the Old Testament 21, Chicago: Baptist Publication Society, p. 5: 'The four Living Creatures... were the common symbols throughout Babylonia for four of their chief deities. The bull colossus, with ox-face, was the symbol of *Marduk*; that with the lion-face was *Nergal*, the god of the underworld and of plague; that of the eagle was *Ninib*, god of the chase and of war; while the human face represented *Nabu*, the announcer or revealer. Each one of these had a long and varied history.... These four were chief in their pantheon at this time, and may be considered as representing all the gods of that empire.'

[276] Greenberg, M. (1983), *Ezekiel 1-20*, New York: Doubleday, p. 57.

these composite beings to underscore their total power. Odell opines that in adapting elements of royal imagery, the Chariot Vision showed that not only was the power of YHWH supreme over earthly power, but it was the only effective power in the lives of the exiles.[277]

The use of redoubling of body parts (four faces of the Living Creatures; multiple eyes on the wheels) was a known feature of Mesopotamian gods from the Kassite period (1,595 BC).[278] The tiaras of Assyrian gods' statues were adorned with many 'eye-stones'.[279] Human bodies of gods often kept an animal head, and this was to distinguish their personality, and 'in some way recall the qualities attributed to the god'.[280] There is a certain irony in Ezekiel using the symbols of the enemy to present to his own people. Petersen says that Ezekiel uses Babylonian imagery to identify the nation that will afflict Israel.[281]

Early Astrology/Astronomy

The documentation of recognisable patterns in the grouping of stars (constellations) originally came from Mesopotamia.[282] There were two reasons for this Mesopotamian interest: firstly, the constellations represented the gods and secondly, they gave a timing to seasons that provided a calendar for farming. These two focuses developed into two traditions: the 'divine' and the 'farm-

---

[277]   Odell, S. (2005), *Ezekiel*, Smyth & Helwys Bible Commentary, Macon: Smyth & Helwys, p. 36

[278]   Mischwesen, A. (1997), 'Philologisch, Mesopotamien,' in Wiggermann, F. A. M., *Reallexikon der Assyriologie (RlA)*, **8**, pp. 222–224; p. 236.

[279]   Vogt, E. (1978), 'Der Sinn des Wortes "Augen" in Ez 1,18 und 10, 12,' *Biblica*, **59**, pp. 93–96.

[280]   Cerny, J. (1957), *Ancient Egyptian religion*, London: Hutchinson's University Library, p. 40.

[281]   Petersen, B. N. (2012), *Ezekiel in Context*, Princeton Theological Monograph Series, Eugene: Wipf & Stock, p. 152.

[282]   Rogers, J. H. (1998), 'Origins of the Ancient Constellations: I. The Mesopotamian Tradition,' *J. Br. Astron. Assoc.*, **108** (1), p. 9.

ing calendar'. The constellations of these two traditions overlapped but in some parts of the sky there were different constellations for the divine and for farming. Celestial figures were developed to represent constellations starting from about 3,200 BC to about 500 BC.[283] The 'divine' tradition identified heraldic animals and divine figures in the constellations for religious purposes. These were transmitted to the Greeks and became the zodiac that we use today. The 'farming calendar' tradition identified rustic workers and farm animals in constellations that served as an annual calendar for farmers.[284]

The faces of the Living Creatures in the initial Chariot Vision of a man, a lion and an ox correspond with the zodiac signs for the lion (Leo), the water-carrier (human) and bull (Taurus). As discussed below, there is evidence that the eagle became an alternative to Scorpio in Babylonia (see section 'The Eagle'). These four zodiac signs are in four quadrants of the zodiac each separated by about 90 degrees of longitude. They thus represent the four quadrants of the heavens and the two solstices (Leo and Aquarius: summer and winter) and two equinoxes (Taurus and Scorpio: vernal and autumnal).

Since the zodiac originated in Babylonia, it is important to look at the origins of the zodiac and of these four figures, in particular:

There were six stages in the development of the constellations and the figures assigned to them in Babylonia:[285]

The early pictographic phase from about 3,200 to 2,100 BC

In this phase, animals were carved on artwork and pottery. Some of these were natural, others were composite anthropomorphic animals with combinations of human and animal characteristics.

---

[283] Rogers, J. H. (1998), *Ibid.*, **108** (1), p. 9.

[284] Rogers, J. H. (1998), *Ibid.*, **108** (1), p. 9..

[285] Rogers, J.H. (1998) 'Origins of the Ancient Constellations: I. The Mesopotamian Tradition,' *J. Br. Astron. Assoc.*, **108** (1), p. 9.

The boundary-stone pictographic phase from about 1,350 to 1,000 BC

Boundary stones were set up to protect the ownership of land that were decorated with symbols of the gods, most corresponding to planets or constellations.

The 'Three Stars Each' phase: after 1,100 BC

The Babylonian sky was divided into three sectors or 'Ways'. Three stars were assigned to each month of the year and were listed on Three Stars Each tablets.

The 'MUL.APIN' phase from 1,100 to 700 BC

This was marked by use of a formal compendium of Babylonian astronomy. MUL.APIN*[286] was the name of the first constellation of the year and the opening words of the compendium.

The Astrometric Diaries phase from about 750 BC to 60 BC

In this phase, astrology and astronomy were mixed together. The Dendera Zodiac (Figure 2) dates from this age.

Transmission of the zodiacal constellations to the Greeks

Since Babylonian astronomy/astrology was part of the *Sitz im Leben* of the Chariot Vision, tracing the origins of the four faces of the Living Creatures in Babylonian astronomy may give insights into what they signify in the Chariot

---

[286]   ✴ (which is transliterated/read as: MUL) is one of many cuneiform 'determiner glyphs'. Here it is used as an ideogram to mean 'constellation' and is made up of three eight-rayed star shaped cuneiform glyphs called 'AN'. AN signifies the concept of 'heaven'. These glyphs can also be used as logograms with a word meaning or as phonograms that can be used to spell out any word phonetically. The phonogram often uses the first syllable of the primary meaning of the ideogram. Several different ideas or meanings can be conveyed by the same ideogram. For example: AN also means 'deity'. The descriptor MUL.APIN denotes the combination of MUL with ✴⊨ (APIN) the glyph for a plough. MUL.APIN ✴ ✴⊨ means 'an astronomy compendium'.

Figure 1: Modern Zodiac.

Figure 2: Dendera zodiac.

Vision. For this exercise therefore, I am going to work backwards through the six phases of development of the constellation figures as outlined above to try and trace their earliest origins.

The most recent, or sixth phase mentioned above is a gradual transmission of the constellations to the Greeks, after the downfall of Babylon in 539 BC. This was after the death of Ezekiel in 570 BC, so this phase does not concern us directly. It has been found however[287] that the two points at which the earth's equator intersect the ecliptic plane*[288] (vernal and autumnal equinoxes), move westward along the ecliptic by 360 degrees per 25,800 years (called 'precession'). Since Babylonian times this has resulted in the earlier appearance of zodiacal signs compared with 3,000 years ago.*[289] We now recognise 13 signs including a new sign, Ophiuchus; whereas the Babylonians recognised 12 (Figure 1).

The astrometric diaries of the fifth phase chronicled historical events and astronomic measurements. Before 475 BC the Babylonians divided the zodiac into twelve equal signs. There was an indication that the Babylonians knew

---

[287] NASA (2018) Image Science Centre at
https://image.gsfc.nasa.gov/poetry/ask/q1795.html.

[288] The imaginary path the sun traces across the sky.

[289] *https://spaceplace.nasa.gov/starfinder2/en/* (about 40 degrees since 2000 BC): constellations have therefore moved about one place forward.

Figure 3: Scorpion-man with wings, lion's feet: 12th century entitlement stone.

Figure 4: Lamassu with lion's feet (Ashurnasirpal II, 883–859 BC).

about precession since measurements from Aries to the point of the spring equinox changed over time.[290] The Dendera zodiac, which was produced by the Egyptians, but shows Babylonian knowledge of the zodiac, was produced in about 50 BC (Figure 2). It shows the twelve signs of the zodiac in addition to other constellation signs recognised by the Egyptians,[291] including the four signs of the equinoxes (Taurus and Scorpio) and solstices (Leo and Aquarius) that were mirrored in the Chariot Vision (these four signs are shaded in Figure 2).

The fourth (MUL.APIN) phase was when the second formal compendium of Babylonian astronomy was written. It existed in tablet form in several copies. The constellation and star lists were a continuation of the third phase ('Three Stars Each') star lists but were improved by accurate observations made around

[290] Rogers, J. H. (1998), 'Origins of the Ancient Constellations: I. The Mesopotamian Tradition,' *J. Br. Astron. Assoc.*, **108** 1, p. 23.

[291] Lull, J., Belmonte, J.A. (2009), 'Constellations of Ancient Egypt,' in Belmonte, J.A., Shaltout, M., *In Search of Cosmic Order: Selected Essays on Egyptian Archaeoastronomy*, Cairo: Supreme Council of Antiquities Press, p. 192.

1000 BC.[292] The first tablet contains most of the zodiac but with farming calendar alternatives for Pisces, Aries and Virgo. The second tablet includes *omina*[293] connected with appearance of the stars, planets, comets and winds. Taurus, Scorpio, Leo and Aquarius are shown here in cuneiform script with their Babylonian transliterated names and the meaning of these names. The Babylonian origins for each of these will be discuss in detail:

𒀯𒄞𒀭𒈾 ^{MUL}GU$_4$.AN.NA The Bull of Heaven; Spring Equinox; Taurus

𒀯𒄊𒄖𒆷 ^{MUL}UR.GU.LA The Lion; Summer Solstice; Leo

𒀯𒄒𒋰 ^{MUL}GIR$_2$.TAB The Cutter/Scorpion; Autumn Equinox; Scorpio

𒀯𒄖𒆷 ^{MUL}GU.LA The Great One;[294] Post-winter solstice; Aquarius

The Great Lion

Although the lion of the modern constellation, Leo, has its origin in the Great Lion of the Babylonian constellation, the Great Lion differs in that it has wings. In the MUL.APIN text the Lion is named as *Latarak*, a minor god.[295] The Great Lion is more a protective figure or a beneficial demon than a god, however. He may be shown holding a whip in ancient art. *Latarak* is frequently twinned with *Lulal* another lion-headed god. The Lion is particularly associated with the goddesses *Inanna* and *Ishtar*. *Inanna* has been called the Lioness of Heaven.[296] *Inanna* had warlike attributes and a lust for war and carnage. Venus is *Inanna*'s sacred planet. There was a close association between

---

[292] Rogers, J. H. (1998) 'Origins of the Ancient Constellations: I. The Mesopotamian Tradition,' *J. Br. Astron. Assoc.*, **108** (1), p. 17.

[293] *Omina* are predictive statements about celestial events. See 'Metaphor and Symbolic Reasoning'.

[294] White, G. (2007), *Babylonian Star-Lore*, London: Solaria, p. 43.

[295] Black, J., Green, A. (1992), *Gods, Demons and Symbols of Ancient Mesopotamia*, London: British Museum Press, p. 116.

[296] The Electronic Text Corpus of Sumerian Literature, Oriental Institute of Oxford University. *A Hymn to Inana as Ninegala*, lines 1–8. At: *http://etcsl.orinst.ox.ac.uk/section4/tr4074.htm* accessed 29/6/2019

Venus and the Lion, the Lion constellation being called Venus' secret place: 'If Venus reaches her secret place: good fortune will come to pass – She reaches the constellation of the Lion.'[297] When Venus appears in the Lion constellation, it portends war. 'If Venus stands inside the Lion in the east: in Elam there will be a battle. If in the west: in Akkad there will be a battle.'[298]

The lion's character is based on three main powers. (a) as the killer of man and livestock and symbolises the carnage of war; (b) the lion often represents the king as a ferocious warrior; (c) the lion also signifies the power of the sun in summer causing drought and death.[299] In terms of astrological predictions, the constellation of the lion symbolised war and destruction of livestock. In Mesopotamia, ravaging lions were common and attacked herds of cattle or sheep. Predictions such as 'If Mars enters the Lion and stands there: downfall of cattle'[300] or 'If the Lion is dark: for three years, lions and wolves will kill people, they will cut off the traffic with the Westland.'[301]

There is a consonance between the qualities of the lion and that of the sun or sun god. An Akkadian design shows lions on the celestial gates, between which the sun rises.[302] The sun god is known as the 'great lion'. The lion represents the sun-god in his aspect as a warrior. He is described with his battle-mace, adorned with three lion's heads and makes the mountains tremble and destroys rebels' lands.[303] The sun is also closely related 'to the concept of the fertility

---

[297]   Reiner, E., Pingree, D. E. (1998), *Babylonian Planetary Omens*, Part 3, Styx, Leiden: Brill, p. 233, omens 13–6.

[298]   White, G. (2007), *Babylonian Star-Lore*, London: Solaria, p. 185.

[299]   White, G. (2007), p. 185.

[300]   Hunger, H. (1992), *Astrological reports to Assyrian Kings: State Archives of Assyria*, Vol. VIII, Helsinki: Helsinki University Press, Report 81.

[301]   Hunger, H. (1992), *Ibid.*, Report 324.

[302]   White, G. (2007), *Ibid.*, p. 187.

[303]   The Electronic Text Corpus of Sumerian Literature, Oriental Institute of Oxford University. *An Adab to Utu for Shulgi*, (lines 12–22)

of the skies where all life originates and the light that fills the heavens.'[304] The lion represented the king, as the lion is the king of beasts. Predictions were made about the king on the basis of the birth of children who had lion-like facial or head appearance.[305]

The Bull of Heaven

The wild bull of Mesopotamia, which is now extinct, was also known as the aurochs. The wild bull was a formidable creature, nearly six feet at the shoulders with huge horns and weighing three quarters of a ton. In ancient Babylonia it was venerated as a symbol of supernatural strength and ferocity. It was only likened to kings, gods or heroes, not to ordinary people.[306] It was related to the European bison and was the ancestor of modern cattle. The wild bull became extinct in 1,627 AD. The bull was a symbol of prosperity 'if the Bull of Heaven's stars are very bright, the offspring of cattle will thrive.'[307] During the fourth millennium BC the bull was associated with several gods and the name, the Bull of Heaven originated as a constellation name. A very early cylinder seal from the Uruk period (4,000 to 3,100 BC) shows the bull as a constellation associated with the goddess *Inanna*.[308] There is an association between the Bull of Heaven and the sun which is seen in the Gilgamesh Epic and the omen literature.[309] An interesting link with the Chariot Vision comes from a pedestal stone[310] which shows a bull's head (a link to the Living Creatures) being venerated. Around the bull's head are several star symbols: the sun (mounted on the bull's

---

[304] White, G. (2007), *Ibid.* p. 188.

[305] White, G. (2007), *Ibid.*, p.186.

[306] White, G. (2007,) *Babylonian Star-Lore*, London: Solaria, p. 84.

[307] Reiner, E., Pingree, D. E. (1981), *Babylonian Planetary Omens*, Part 2, Malibu: Undena, Text XV, line 30, p. 75.

[308] Glassner, J. J. (2003), *The Invention of Cuneiform*, Baltimore: Johns Hopkins University Press, p. 175.

[309] White, G. (2007), *Ibid.*, p. 86.

[310] Dalley, S. (1986), 'The God *Salmu* and the Winged Disk,' *Iraq*, **48**, pp. 85–101.

head) moon, and Venus. The bull represents the cool wet season as opposed to the hot summer season represented by the lion. Babylonian seal designs often showed them fighting – a metaphor for the changeability of the weather. The Bull of Heaven's constellation shifted due to precession from the spring equinox to the second month of the Babylonian year,*[311] when rains diminish in early summer. This resulted in a change in the attributes of the Bull in myth stories, from one associated with spring and birth to one of a raging bull, that destroys pastures and dries rivers.[312]

The Great One (The Babylonian precursor of Aquarius)

The constellation of *Gula* or 'the Great One' is of a standing man holding a jar in each hand from which water is overflowing. He is pictured as standing on the earth and the overflowing water symbolises rain. This constellation represents the rain that falls in the winter and swells the rivers.[313] The Great One is closely associated with *Enki*, the god of wisdom and water. *Enki* is one of the three most respected gods in Mesopotamia: *Anu* (the most ancient god and father of all gods), *Enlil* (the god of farmers and the leader of the Babylonian pantheon) and *Enki*. These three gods constructed heaven and earth, divided the heavenly paths and drew the constellations and created day, night, month and year.[314] *Enki*, or the Lord of Springs, (*Enki-Ea*) is shown with water overflowing and may be seated in a square enclosure that represents the Abyss. During the second millennium symbolism of water and winter constellations were associated with *Enki*. *Enki* was loved and respected as a wise and benevolent god who had invented the arts and crafts of civilisation. He makes rain fall

---

[311] The Babylonian new year started at the sighting of the moon at the spring equinox.

[312] White, G. (2007), *Ibid.*, p. 89.

[313] Porada, E. (1987), *On the Origins of Aquarius: Festschrift für Reiner*, Ann Arbor: American Oriental Society, pp. 279–91.

[314] Horowitz, W. (1998), *Mesopotamian Cosmic Geography*, Winona Lake: Eisenbrauns, pp. 146–7.

from heaven and in one poem he is actually identified with the waters.[315] *Enki* brings fertility and progeny to creatures and life to cultivated fields and his waters also govern human reproduction.[316] *Enki* has the title 'Father' (*Aya*) and the filling of waters with fertility also have a sexual symbolism.[317]

The Scorpion

Scorpio is reviewed here because Scorpio is the traditional constellation of the autumnal equinox and would be the expected representation of the fourth quadrant of the heavens in the faces of the Living Creatures of the Chariot Vision. However, there is evidence (reviewed below under 'The Eagle') that the scorpion and the eagle were interchangeable and that the scorpion may have been replaced by the eagle in the Lamassu. The scorpion's sting, pincers and armoured body symbolise the martial prowess of the king. One omen says: 'If the Scorpion comes close to the front of the moon and stands there: the reign of the king will become long; an enemy will attack but will fall.'[318] In warlike characteristics the scorpion is linked to the lion and the bull: 'its pincers extended like the horns of a wild bull and its tail raised up like that of a raging lion.'[319] Akkadian period artwork shows a composite scorpion-man with wings and lions feet (Figure 3), the front end of which is reminiscent of the Lamassu. In the stellar calendar, the scorpion's rising just before the winter

---

[315] The Electronic Text Corpus of Sumerian Literature, Oriental Institute of Oxford University, *Enki and the World Order*, lines 89–99. At: *http://etcsl.orinst.ox.ac.uk/section4/tr4074.htm* accessed 29/6/2019.

[316] White, G. (2007), *Babylonian Star-Lore*, London: Solaria, p. 160.

[317] The Electronic Text Corpus of Sumerian Literature. Oriental Institute of Oxford University, *Enki and the World Order*, lines 250–266. At: *http://etcsl.orinst.ox.ac.uk/section4/tr4074.htm* accessed 29/6/2019.

[318] Hunger, H. (1992), *Astrological reports to Assyrian Kings: State Archives of Assyria*, Vol. VIII, Helsinki: Helsinki University Press, Report 430.

[319] White, G. (2007), *Ibid.*, p. 232. retranslated from Freedman 1998–2006. If a city is set on Height. *Shumma Alu*, University of Pennsylvania Museum, p. 159, tablet 31, text E, lines r5–12.

solstice marks the descent of the sun into the darkness of the underworld, symbolising the 'death' of the sun and the end of the year.[320]

The Scorpion is linked to the goddess *Ishara* who was worshipped throughout the Ancient Near East. Several titles were attributed to her including goddess of love, '*Ishara* of the ocean', and queen of the inhabited world. Her main attribute however was goddess of war and victory.[321] Astrologically the time of scorpion's actual first appearance, when compared to the ideal time of the constellation's rising, is a way of predicting the fortunes of war. 'The Scorpion rises in month 8: if this star rises early: the king will go about proudly; he will subdue his enemies. If this star rises late: the kings of all lands will start hostilities against the king.'[322]

The Eagle

The Eagle is the Babylonian equivalent of the Greek constellation Ἀετός. The constellation rises in the month before the winter solstice, at the end of the stellar calendar, like Scorpio but later. This gave a death function to the constellation since it was the month that traditionally, the souls of the dead ascended to heaven and joined the stars of the Milky Way where the ancestors dwell. Eagles were involved in ferrying the souls of the dead.[323] The eagle has been portrayed standing on a pillar as the primary representation of the warrior god *Ninurta*. An eagle with a lion's head or a weapon with the head of an eagle were also used.[324] The sacred symbols of the warrior god *Zababa*[325] are

---

320 White, G. (2007), *Ibid.*, p. 232.

321 White, G. (2007), *Babylonian Star-Lore*, London: Solaria, p. 234.

322 Hunger, H. (1992), *Astrological reports to Assyrian Kings: State Archives of Assyria*, Vol. VIII, Helsinki: Helsinki University Press, Report 502, lines 11–4; and report 219, lines 1–6.

323 White, G. (2007), *Ibid.*, p. 126.

324 Baigent, M. (2005), *Astrology in Ancient Mesopotamia*, Rochester: Bear & Co., p. 137.

325 Sometimes identified with Ninurta.

two scimitars, one representing a lion and the other an eagle.[326] The symbol of an eagle was regularly found on Babylonian entitlement stones, guarding property limits, together with certain weapons.[327]

The eagle may have become equivalent to the scorpion in the zodiac because the eagle was regarded on par with the lion as a war-like symbol and it has been portrayed as a solar symbol. In comparison with the scorpion, which tends to lie low and strike from hidden corners without warning, the eagle also strikes without warning but the eagle soars, is proud and free and has an 'eagle eye'. Possibly these characteristics won over those of the scorpion for the Lamassu, as being more becoming of a king: being all-seeing and lofty rather than underhand and secretive.[328] Also, the eagle's wings are more aesthetically pleasing than the scorpion's sting.*[329] (Figures 3, 4) These may also be reasons why the eagle would have been portrayed in the Chariot Vision rather than the scorpion. These four signs (human, lion, eagle and bull) of presumed zodiacal origin, are represented in the faces of the Living Creatures of the Chariot Vision.[330]

The Third Phase was that of the 'Three Stars Each' tablets, starting 1,100 BC. This was the earliest recorded Babylonian star system. It was also called the '36 stars system' as it contained 36 stars, three for each month. The stars

---

[326] White, G. (2007), *Ibid.*, p. 305.

[327] Baigent, M. (2005), *Ibid.*, p.137.

[328] Farrer, A. (2005), *The Revelation of St. John Divine*, Eugene: Wipf & Stock, p. 91. [original publication (1964) Oxford: Oxford University Press] states that the Scorpion is the Eagle's zodiacal equivalent as they had equivalent heliacal rising times.

[329] The two pairs of wings that the Living Creatures had in the Chariot Vision do not appear to be related to the eagle face, since the Living Creatures were also considered a type of angel and were called cherubim (Ez 10.1), and they covered their bodies with two of the wings, (Ez 1:11) like the seraphim in Is 6:2.

[330] The identification of the Living Creatures with these four zodiacal signs is supported by Tim Hegedus (*Early Christianity and Ancient Astrology*, Patristic Studies Volume 6, New York: Peter Lang 2007, p. 234) as the creatures surround the heavens; It is also supported by Farrer, A. (2005), *Ibid.*, 91–92, who specifically identifies the human face of the creatures with Aquarius and the eagle face with Aquila.

were divided between the three ways of the heavens, based on the creation epic *Enuma Elish* (described in the Second Phase below).[331] The text lists each of these stars and gives their position, rising and setting and significance for mythology and agriculture. The civil year in Babylonia was determined by lunar months, and the new year started at the sighting of the moon at the spring equinox.*[332] The year might therefore have 12 or 13 months depending on the relationship of the lunar to the solar year. In Egypt the heliacal rising*[333] of Sirius marked the start of the new year. It also heralded the flood of the Nile. Around 3,000 BC in Mesopotamia the heliacal rising of MUL.APIN ('Plough' or the modern 'Triangulum' constellation) heralded the start of spring ploughing. It is interesting to note that the only constellations represented in the zodiac at this time were Leo, Scorpius, Aquarius and the Pleiades[334] (Pleiades is within the modern constellation of Taurus). These four constellations (with the eagle replacing the scorpion) are those that are represented in the four faces of the Living Creatures of the Chariot Vision.*[335]

The Second Phase was the phase of boundary stone pictographs. During the Kassite dynasty (1,531BC–1,155 BC) boundary stones were set up which guarded property boundaries. The stones were royal charters which called upon the gods to bear witness and protect the land boundary. They were decorated with the symbols of gods which corresponded to the planets and constellations.[336] In the second millennium BC there are clear records of divine icons which

---

[331] Rogers, J. H. (1998), 'Origins of the Ancient Constellations: I. The Mesopotamian Tradition,' *J. Br. Astron. Assoc.*, **108** (1), p. 16.

[332] The method of timing of Easter is based on the Babylonian calendar.

[333] The heliacal rising or star rise of a star occurs annually when it briefly becomes visible above the eastern horizon at dawn just before sunrise, after a period of less than a year when it had not been visible.

[334] Rogers, J. H. (1998), 'Origins of the Ancient Constellations: I. The Mesopotamian Tradition,' *J. Br. Astron. Assoc.*, **108** (1), p. 17.

[335] As noted above, Aquila, the Eagle and Scorpio became interchangeable at a later date.

[336] Rogers, J. H. (1998), *Ibid.*, **108** (1), p. 12.

became the zodiacal constellations of Taurus, Leo, Scorpius and Aquarius in addition to others. Rogers illustrates a late Kassite boundary stone which shows a winged lion striding on the back of a serpent, the same posture that was shown for Leo in the Dendera zodiac,[337] (Figure 2). This suggests that the anthropomorphic boundary stone symbols were the likely origin of the zodiac constellations.

An important text of this time was the great 'epic of creation'.[338] It tells the story that the ocean was without form and ruled by the dragon *Tiamat*. He summoned several creatures to help him including several anthropomorphic monsters seen on boundary stones, including the Great Lion, and the Scorpion-man. *Marduk* (Jupiter) was the hero who tore *Tiamat* into two parts, making one into the heavens with its constellations, and the other into the earth.

> He constructed stations for the great gods,
> Fixing their astral likenesses as constellations.
> He determined the year by designating the zones:
> He set up three constellations for each of the twelve months,
> After defining the days of the year [by means] of (heavenly) figures,
> He founded the station of *Nebiru* (Jupiter) to determine their heavenly bands,
> That none might transgress or fall short.[339]

The very early first phase of early pictography starting about 3,200 BC included cylinder-seal impressions that showed scenes of animals and gods with the symbols for the Sun, Moon and Venus. From these scenes, standard symbols for the gods which were shown in the sky as well, evolved as the zodiacal iconography.[340]

---

337 Rogers, J. H. (1998), *Ibid.*, p. 12.

338 Pritchard, J. B. (2011), *The Ancient Near East: An Anthology of Texts and Pictures*, Princeton: Princeton University Press, p. 28.

340 Mischwesen, A. (1997), 'Philologisch. Mesopotamien,' in Wiggermann, F. A. M. *Reallexikon der Assyriologie (RlA)*, **8**, pp. 222–224; p. 229.

The Lamassu

The huge ſtatues of composite anthropomorphic creatures that were set as palace guardians in Babylonia are of intereſt because the composite of animals was similar to the living creature faces in the Chariot Vision. A Lamassu composed of four animals (human, bull, presumed eagle and lion) (figure 4) was firſt sculpted for Ashurnasirpal II (883–859 BC) for his palace which he inaugurated in 879 BC.[341] Lamassu at the north-weſt palace of Ashurnasirpal II[342] had human arms in front.*[343] Winged animals with human heads were a very ancient anthropomorphic figure and seen as early as 3,000 BC in Babylonia. The earlieſt example of a Lamassu was from the reign of Tiglath-Pileser I (1,114–1,076 BC), a winged bull with a human head,[344] and many Lamassus have this combination.

Two queſtions about the Lamassu are (a) whether the wings are meant to be those of an eagle, and if so, (b) if they have a zodiacal or aſtrological significance. The wings are not recognisable as an eagle's but in the context of the Palace Guardian role of the Lamassu, it is quite possible that they represented an eagle: images of the warrior god and city ruler *Ninurta**[345] were often carved with an eagle together with weapons. The eagle is therefore an eminent candidate for a guardian icon. The development of the Lamassu to include lions' feet by Ashurnasirpal could therefore to have been a portrayal of the four quadrants of the heavens in the zodiac.

---

[341] *https://www.metmuseum.org/toah/works-of-art/32.143.2/*.

[342] *https://www.abc.net.au/news/2016-11-14/a-lamassu-at-the-north-weſt-palace-of-ashurnasirpal-ii-in-nimrud/8022022*

[343] The Living Creatures in Ez 1:8 had human hands.

[344] Danrey, V. (2004), 'Winged Human-Headed Bulls of Nineveh: Genesis of an Iconographic Motif,' *Iraq*, **66**, pp. 133-139.

[345] Ninurta was known as 'the hero of Enlil' and was famous for subduing the rebel lands around Mesopotamia. See White, G. (2007), *Babylonian Star-Lore*, London: Solaria, p.73.

For Babylonians, the zodiac was used for two main reasons: for religious purposes and to mark the seasons as an objective method of time-keeping, as a farming calendar.[346] The king associated his power with that of the gods as portrayed in the named constellations of stars and zodiac, in the boundary markers and also in statues like the Lamassu. The four main markers of the zodiac were for the equinoxes and the solstices. These privileged places in the zodiac were given to the bull, the lion, the scorpion (or eagle) and the great one who brings water (later Aquarius). The fact that the eagle was easily substitutable for the scorpion suggests that the significance of the zodiac animal for the king's perceived power was more important than the exact location of a particular constellation. The Chariot Vision, then may have utilised the extant icons of power, to portray the Glory of God.

The identification of the Living Creatures with four key zodiac constellations gives them a cosmic significance. The link of the faces to the zodiac brings a rich perspective of the *Sitz im Leben* of Ezekiel to their assigned character-forces which are summarised in Appendix 12. Although the Creatures are linked through the names of these four constellations, the locations of the constellations change over time by precession, and there is doubt about whether some of the constellations we see are the same as those named in Mesopotamian times.*[347] The real link between the cosmos and the Creatures is not therefore with the zodiac constellations themselves, but with the four 'seasonal' effects of the cosmos on the seasons, the winds and the earth. The faces of the Creatures therefore link to the cosmos through annual seasonal changes.

Cook sees the Living Creatures as key stabilizing elements in the structure of the cosmos, reflecting Ezekiel's gradations of holiness extending through the

---

[346] Rogers, J. H. (1998), 'Origins of the Ancient Constellations: I. The Mesopotamian Tradition,' *J. Br. Astron. Assoc.*, **108** (1), p. 9.

[347] For example, the Plough which heralded spring farming does not have a documented rising or setting and is actually a circumpolar star. White, G. (2007), *Babylonian Star-Lore*, London: Solaria, p. 215.

cosmos both vertically and horizontally.[348] Eichrodt says that the four Living Creatures are representatives of the four corners of the earth because they carry the firmament or vault which the Creator set up (Ge 1:6) to separate the earthly and heavenly waters.[349] Irenaeus of Lyons was the first to see a connection between the four creatures of Ezekiel with the four main zones of the world and the four principal winds.[350]

Anthropomorphism and Choice of Zodiac Creatures

The choice of animal or anthropomorphic creature to name constellations was based more on the preconceived needs of the 'divine' or 'farming' systems of the zodiac than any obvious similarity to the star pattern seen in the heavens and arose out of the Babylonian pictographic tradition. These named constellations acquired a characteristic symbolic nature, often depending on the time of their 'annual rising'.*[351] This symbolic nature was attributed to the constellations by means of the chosen animal that represented them.

There are two levels of anthropomorphism in the zodiac signs to which the Living Creatures faces relate. Firstly, the specific choice of animal is an anthropomorphism of different characteristics that the Mesopotamians held to be 'divine'. Shaman et al's view[352] is that the structure of individuals' anthropomorphic concept of God, can be divided into three domains: the phys-

---

[348] Cook, S.L. (2004), 'Cosmos, *Kabod* and Cherub: Ontological and Epistemological Hierarchy in Ezekiel' in, *Ezekiel's Hierarchical World: Wrestling with a Tiered Reality*, Patton, C.L. and Cook, S. L. (eds.), Atlanta: Society of Biblical Literature, pp. 179–198; p. 184.

[349] Eichrodt, W. (1970), *Ezekiel*, London: SCM Press, p. 58.

[350] Irenaeus, *Adversus haereses*, III.11.7.

[351] Every year the date of the first appearance of a constellation when it briefly becomes visible above the eastern horizon just before sunrise is also called the 'annual rising' – see White, G. (2007), *Babylonian Star-Lore*, London: Solaria, p. 15.

[352] Shaman, N. R. *et al.* (2018), 'Dimensional Structure of and Variation in Anthropomorphic Concepts of God,' *Front. Psychol.*, **9**, p. 1425.

ical, biological, and psychological. The physical relates to the animal chosen, the biological is the relationship of the chosen animal to nature or time of year and the psychological is the perceived divine trait that is evoked by the animal. There is evidence that we apply human-like traits to non-human entities, including supernatural beings, through an innate process. Secondly, the four zodiac animals chosen, likely express an anthropomorphism of aspects of the mind of man. Anthropomorphism potentially relies on a process of imagining others' minds and mental states[353] and is a projection of human characteristics onto non-human beings. Shepard maintains that 'symbolic images of animals ... enable humans to objectify qualities and traits'.[354] In terms of the Living Creatures, it is likely that the four quadrants of the heavens, that related to key seasons of life, were marked by an animal whose anthropomorphic meaning enhanced the king's prestige and standing.

Anthropomorphising is a cognitive process,[355] and has parallels with the metaphorization of celestial *omina* in early Babylonia (See 'Metaphor and Symbolic Reasoning'). Biblical anthropomorphism is an attempted metaphoric enlightenment about divine characteristics.[356] The above analysis of the Mesopotamian origins of the faces of the Living Creatures shows that these were intimately tied to early myth and are an example of early anthropomorphosis. Since ancient myth and *omina* contain fundamentals of pre-science, (See sec-

---

[353]   Severson, R., Woodard, S. R., (2018), 'Imagining Others' Minds: The Positive Relation Between Children's Role Play and Anthropomorphism,' *Front. Psychol.*, **9**, article 2140; An important caveat is that one of these four zodiac signs is represented by a human figure ('The Great One' or *Aquarius* see below). Although 'anthropomorphism' does not apply to a human, the Great One was linked to the god *Enki*, and so was seen as a divine person. A similar anthropomorphic process seems to have been involved in the choice of this figure, which is discussed further below.

[354]   Shepard, P. (1978), *Thinking Animals*, New York: Viking, p. 247.

[355]   Airenti G, *et al.* (2019), 'Editorial: The Cognitive Underpinnings of Anthropomorphism,' *Front. Psychol.*, 10, article 1539, doi: 10.3389/fpsyg.2019.01539.

[356]   Howell, B. (2014), *In the Eyes of God: A Metaphorical Approach to Biblical Anthropomorphic Language*, Cambridge: James Clarke, p. 241.

tion 7.2: Review of Babylonian Metaphor), the original process of selecting the Creatures could also have been a form of pre-scientific thought.

Epley et al state that 'anthropomorphism is multiply determined, involving both cognitive and motivational determinants... a primary determinant of anthropomorphism is therefore the elicitation of agent knowledge itself. Knowledge about humans in general, or about the self in particular, is likely to serve as the basis'.[357] The four faces that subsequently made up the Living Creatures were therefore not selected randomly, when the constellations were first named, but probably through a cognitive process, (possibly subconscious).

Choice of characteristic traits of four creatures

The Christian interpretation of the characteristics of the four Creatures that originated from Irenaeus was that they represented the four gospels. This may have been influenced by the four living creatures around the throne in Re 4:7, but which creature was assigned to which evangelist depended on choosing a particular aspect of the animal. Gregory following Irenaeus saw four aspects of Christ in the Creatures: the human face as his incarnation, the calf as sacrificial and priest, the lion as his imperial aspect and the eagle his spirit. These interpretations were an individual paring of an animal characteristic with an important feature of the gospel. In most of these, the lion was imperial or brave; the calf sacrificial; the eagle spiritual and the man the image of God (Appendix 12 under Scripture column).

A second interpretation that Jerome mentions that may have come to him through Origen aligns three of the Creatures with the Platonic tripartite division of the soul (see subsections 3.4 and 3.19.2). This gives to the man reason, reflection, mind and counsel; to the calf lust, excess and the desire for all pleasures; to the lion fierceness, anger and violence but the eagle is not included.

---

[357] Epley, N. *et al.*, (2007), 'On Seeing Human: A Three-Factor Theory of Anthropomorphism,' *Psychological Review*, **114** (4), pp. 864–88; p. 866.

Kries[246] sees the eagle as the pneuma of Plato and links this with Origen's 'spirit' or conscience. Jerome too, sees the eagle as conscience hovering above and correcting the other three Creatures. This interpretation gives insight into the character forces of the Creatures, but it is not a good fit with the Chariot Vision in that the four Creatures are not equal in their input as the eagle was not represented in Plato's tripartite division. The characteristics given to the Creatures however, fits well with other sources.

Babylonian started assigning character traits assigned to zodiac creatures around 1,100 BC. Despite this, similar traits for the four Creatures emerge. The conflict between the bull (cool wet spring) and the lion (hot summer season) echoes the juxtaposition of the ox and the lion in the axes of the Living Creatures (Figure 6). In the Babylonian tradition the eagle was linked to a spiritual role, the man figure with wisdom and fertility. This links to a juxtaposition in the Living Creatures between the rational human and the ethereal eagle.

Appendix 12 reviews the metaphoric traits that have been associated with the four animals of the Living Creatures, in Babylonian, Biblical and modern times. Despite the passage of time, certain core human characteristics stand out as a typical characterisation of each of the four. In the choice of the descriptive titles for the four creatures, it was important that these three time periods were represented: the Babylonian juxtapositions of seasons, the bible characteristics of the gospels, and descriptors that would be understandable by modern man. Descriptive markers that encapsulate these character-forces*[358] are: rationality for the human; perspicacity for the eagle; authoritativeness for

---

[358] See Appendix 12 for a list of metaphoric traits of the Living Creatures in Babylonian, Biblical and modern times. Although frequently translated as 'bull', the Hebrew word used in Ez 1:10 is *šôr* שׁוֹר which is usually translated as ox but can mean a bull, or cattle (cow or ox). Strong's Concordance: 7794; Strong, J. (2012), *Hebrew Dictionary of the Bible,* London: BN Publishing. Since *šôr* includes all cattle, the traits of the cow and ox are used since these are more representative than those of the bull.

the lion; nurturance for cattle.[*359, *360] The movements of the Chariot will now be examined in the light of these traits.

Fixed Position of the Living Creatures

The Living Creatures' movements were directed by the spirit.[361] The Creatures were able to move in any direction without turning their bodies because they had four faces. The position of the faces in relation to each other were fixed: the front faces of each figure were human; those to the right, a lion; to the left, a bull; and to the rear, an eagle (Ez 1:10). There were therefore twelve faces in all, four of each animal (Figure 5).

This fixed relationship of the four creatures as the chariot moves, suggests that the pattern of the character-forces of the faces could be an analogy of how mankind confronts new situations. Since the Chariot moved in straight lines, the twelve faces allowed the human face to always be in front. This fixed relationship also suggests that the pattern of faces could be a metaphor for the mechanism of decision making.[*362]

---

359 Cattle are domesticated ruminant animals: cows or oxen.

360 Rationality is the quality of being rational or logical; perspicacity is the quality of having a ready insight into things; Authoritativeness is the quality of possessing authority; Nurturance is the quality of readily providing affectionate care and attention.

361 Block, D. I. (1989), 'The Prophet of the Spirit: The Use of RWH in the Book of Ezekiel,' *JETS*, **32**, pp. 27–49; p.36.

362 Block, D. I. (1997), *The Book of Ezekiel Chapters 1–24*, Grand Rapids: Eerdmans, p. 96; Since the faces have fixed positions as the Chariot moves forward the four faces can be grouped according to these traits into more masculine (rationality, authoritativeness) at the front and right and more feminine (perspicacity, nurturance) at the left and back; The second description of the faces of the cherubim in Ez 10:14 gives a different sequence of the faces to those of the Living Creatures in Ez 1:10 and replaces the bull with the face of a cherub. The sequence of human-lion-eagle-bull changes to cherub-human-lion-eagle. Block says that although the cherub replaces the bull, the sequence of the faces is the same as if looked at from the East, Ezekiel's vantage point. He would see the bull/cherub first then human, lion and eagle, Block, D. I. (1997) *Ibid.*, p. 325. The question as to why the bull was replaced and how is the cherub face to be perceived? The Babylonian

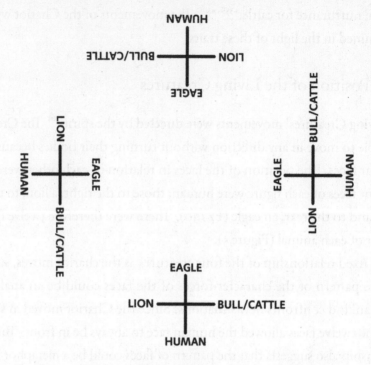

Figure 5: Positions of the twelve faces of the Living Creatures.

The Psychology of Decision Making

In terms of their character-forces, two axes of psychological action can be discerned in the Living Creatures: an anteroposterior or cognitive axis and a lateral axis at right angles to this which is an emotional/affective axis (Figure 6).[363]

In the anteroposterior axis: rationality and perspicacity (insight and intuition) are both cognitive traits that are used in gaining new knowledge. In

---

Talmud says that the bull was replaced because it was a symbol of Israelite sin in the desert (Ex 32:4). The cherub's face was perceived as being small as the Aramaic '*kĕrabya*' meaning 'like a child' is like the word for cherub.

[363] Figure adapted from Block, D. I. (1998), *The Book of Ezekiel Chapters 25–48*, Grand Rapids: Eerdmans, p. 325.

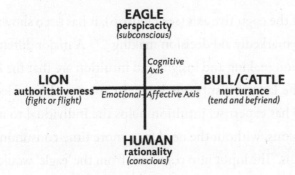

Figure 6: Psychological axes of the Living Creatures.

the lateral axis, nurturance which is an emotionally based trait, is opposite authoritativeness, which is an affective attitude.*[364] This 'emotional' axis incorporates the two main behavioural responses to stress in males (fight or flight for the lion) and females (tend and befriend for cattle).[365] These two axes, the cognitive and the emotional are the two major psychological systems used in decision making.[366] The balancing of authoritativeness against affectionate care (or nurturance) is a very fundamental balance in decision making and conscience,[367] which is represented in the 'emotional' axis of the Living Creatures.*[368]

---

[364] An affective attitude is an emotion generated in response to a person, object, or event.

[365] Taylor, E. *et al.* (2000), 'Biobehavioral responses to stress in females: tend-and-befriend, not fight-or-flight,' *Psychol. Rev.*, **107**, pp. 411–29.

[366] Damasio, A. R. (2005), *Descartes' Error: Emotion, Reason and the Human Brain*, New York: Harper Collins, p. 192.

[367] This conflict is reflected in a Mesopotamian Cylinder Seal (c. 3rd century BC) depiction of the lion-bull conflict between spring and summer weather: Collon, D. (1987), *First Impressions: Cylinder Seals of the Ancient Near East*, London: British Museum Press, Figure 933, p. 192.

[368] This balance is also reflected in the switch of seasons and of divine mood between the first and last Visions of the Chariot as discussed below 'Movements of the Chariot'.

In terms of the cognitive axis (see Figure 6), it has been shown that insight and intuition markedly aid decision making.[369] A major difference between rational decision making and insight and intuition are that the former is conscious, and the latter are subconscious. In an area of knowledge in which an individual has expertise, intuition helps the individual to make fast and effective decisions, without the need for a more time-consuming, conscious, rational analysis. The input into cognition from the 'eagle' would therefore be subconscious. Origen says that it is through the eagle that the Spirit exerts its effect.[370] Even in the fourth century, Jerome saw the four Creatures as decision making forces in humans, with the man, lion and calf making decisions but overlooked by the eagle who has an innate drive towards the good, 'which does not mingle with the other three but corrects them when they go astray'.[371]

Living Creatures and the Spirit 'Each moved straight ahead'

In Ezekiel 1 and 10, in the description of the Chariot Vision there is an extraordinary emphasis on the rectitude of the chariot movements. There are seven mentions of either 'each moved straight ahead: *'îš 'el-'eber pānaw yēlēkû* (איש אל־עבר פניו ילכו) in Ez 1:9 and 1:12, or 'without turning as they went': *lô-yissabbû bĕlektān* (לא־יסבו בלכתן) in Ez 1:9,1:12,1:17, also twice in 10:11 – (where the appropriate masculine form *bĕlektām* (בלכתם) is used. This repetitive mention suggests that these phrases are important and hints that there is a specific

---

[369] Intuition and insight are largely subconscious processes: when a person arrives at an idea or a decision not by analytically inferring the solution, if the correct solution is 'sensed' without being able to give reasons for it this is 'intuition'; if the solution is realized all of a sudden without being able to report on the solution process, this is 'insight', Cf. Zander, T. *et al.* (2016), 'Intuition and insight: Two Processes that Build on Each Other or Fundamentally differ?,' *Front. Psychol,* 7, article 1395.

[370] Origen (2010), *Homilies 1–14 on Ezekiel,* trans Sheck, T. P., New York: Newman Press, p. 45.

[371] *Commentary on Ezekiel,* Ancient Christian Writers series, The Newman Press: New York, 2017, p. 22; The word 'synderesis' (συντήρησις), was first coined by St Jerome.

underlying meaning. The Living Creatures have their own will, for 'wherever the spirit would go, they went':*[372] *'el-ašer yihyeh šamma harûah laleket yēlēkû* (אל אשר יהיה־שמה הרוח ללכת ילכו) in Ez 1:12, but they obeyed the spirit's wishes.*[373] Going straight and not turning, would need a careful coordinated effort, each of the four creatures cooperating fully. Straight movement therefore needs not only perfect obedience to the spirit, but precise harmony and self-control in the creatures.

The Living Creatures have a resemblance to man: *děmût 'ādām lāhenna* (דמות אדאם להנה) in Ez 1:5 with a man's face in front and they had hands (Ez 1:10). Despite having four faces of different animals, the composite creatures had a human body and Maimonides gives them intellect.*[374] The Creatures have a combination of human and angelic anatomy which suggests, like cherubim, they represent an interaction of the divine and human.[375] To reinforce this,

---

[372] Like the English verb 'to go' the Hebrew הלך meaning to go, come or walk, is almost always used intransitively (*Strong's Concordance,* 1980). ללכת therefore does not imply a transitive 'sending' and ילכו would therefore not denote a passive 'transitive' movement, but an action initiated by the Creatures.

[373] Odell says the fact that the living beings are directed by the spirit shows that the living beings are part of the created order and that they move in harmony because God keeps all in perfect control and have become servants to the enthroned glory. Odell, M. (2005), *Ezekiel,* Smyth & Helwys Bible Commentary, Macon: Smyth & Helwys, p. 32.

[374] Maimonides interpretation (of Ezekiel's Living Creatures) is that since the four faces are variants of human faces (all four creatures had a man's bodily form/likeness), they are connected by the fact that they all have intellect, see Davies, D. (2011), *Method and Metaphysics in Maimonides Guide for the Perplexed,* Oxford: Oxford University Press, p. 118; If, as discussed below, the creatures have human reason, then, according to Aquinas, (*Summa Theologiae,* II 2:49) they would not be angels; Eichrodt says that Ezekiel stresses that the Living Creatures have a resemblance to man with a man's face in front and they have hands (Ezek 1:10) see Eichrodt, W. (1970), *Ezekiel,* London: SCM Press, p. 55.

[375] Block, D. I. (1997), *The Book of Ezekiel Chapters 1-24,* Grand Rapids: Eerdmans, p. 319; Launderville says that the cherub marks out a special location where the divine and human intermingle in a measure beyond that of typical human experience; see Launderville, D. (2003), 'Ezekiel's Cherub: A Promising Symbol or a Dangerous Idol?,' *Catholic Biblical Quarterly,* **65** pp. 165–183, here p. 168. The cherub is able to balance competing religious, and social forces. Launderville (2003), *Ibid.,* p. 166.

the Creatures have two types of movement: a vertical movement, ostensibly by using their wings and a horizontal earthly movement transmitted by the wheels. They mounted up from the earth using their wings (Ez 10:16) but the straight movement without turning relates to their progression across the earth, measured by the wheels. The fact that the Creatures were able to make this earthly progression without turning relates to a balancing of the wills of the four animals represented by each face. This coordination is possible through the spirit, *hārûaḥ* (הרוח) The force that causes the movement of the Creatures is a form of divine intelligence and power.[376] The Creatures may be said to have a 'spiritually-dependent directional potency'.

The Creatures' horizontal movement is a sudden darting lightning-like movement. Launderville contrasts this with the statue-like pose of the Creatures with which these darting movement are in tension.[377] The only movement that the Creatures appear to make of themselves is of their wings when they move vertically. The statue-like bearing of their human resemblance,[378] suggests that the earthly will of the Creatures has been totally given over to the spirit but they still exercise a heavenly will. To understand the movements of the Creatures therefore, it is important to explore these heavenly and earthly interactions further through the faces of the Creatures.

Moving Straight Ahead and Right Reason

The analysis above showed that the movements of the Living Creatures 'each moved straight' likely has a specific message. The movements are not arbitrary but based on an interplay of character-forces. This suggests that their movements are reason-based and that 'moving straight ahead', may be an analogy

---

376 Launderville, D.F. (2007), *Spirit and Reason*, Waco: Baylor University Press, p. 58.

377 Launderville, D.L. (2007), *Ibid.*, p. 58.

378 Eichrodt, W. (1970), *Ezekiel*, London: SCM Press, p. 55.

for 'right reason'.[379] St Augustine, supported by Aquinas, within 'right reason' distinguished 'inferior reason' applied in acquiring science (knowledge) and 'superior reason' in attempting to acquire wisdom.[380] Although the rational underpinning of the Creatures' movements supports inferior reason, the presence of the spirit, and the docility of the Creatures to the spirit, suggests that the Creatures are also a model for superior reason and the attainment of wisdom.

Each Creature has to function within a division of power among the four faces of its head.[381] The Creatures move in harmony and 'moved straight', because the spirit keeps all in perfect control. The straight-leggedness of the Creatures signifies they characteristically stood upright and did not kneel in the divine presence,[382] which could be a sign of pride or defiance to divine rule. The four Creatures are therefore in balance between their inherent positive and negative tendencies. Odell sees the living beings as demons, who have come under control.[383] Launderville says that the spirit suffused the Living Creatures like it suffused Ezekiel.[384] The will of the creatures, therefore, has become totally docile to the spirit.

---

379 Fagothey, A. (1972), *Right and Reason,* 5th edition, St Louis: Mosby, p. 102, defines 'right reason' as reason rationally exercised, consistent with itself and faithful to its own law and function; Pullicino, P. (2021), 'Each moved straight ahead: The Subjection of Will of the Living Creatures to the Spirit (Ezekiel 1:12),' *Polish Journal of Biblical Research,* **19—20,** pp.105–122.

380 St Augustine, *De Civitate Dei,* 40 27:2; St Thomas Aquinas *Summa Theologiae,* I.79:9.

381 Launderville, D. (2004), 'Ezekiel's Throne-Chariot Vision: Spiritualizing the Model of Divine Royal Rule,' *Catholic Biblical Quarterly,* **66,** pp. 361–377; p. 366.

382 Mackay, J. L. (2018), *Ezekiel: A Mentor Commentary,* Volume 1, Chapters 1–24, Fearn: Christian Focus Publications, p. 88.

383 Odell, M. S. (2005), *Ezekiel,* Smyth & Helwys Bible Commentary, Macon: Smyth & Helwys, p. 27; The four creatures could therefore represent the negative aspects of the animal they portray, which can be equated with four major sources/sins of human pride: intellectual, power, possessions and spiritual.

384 Launderville, D. F. (2007), *Spirit and Reason,* Waco: Baylor University Press, p. 374.

Moral Decision Making

Recent psychology research stresses that emotional input is essential for learning as well as important in decision making.[385] Emotion may be an important reason for irrational behaviours in fairness-related decision making.[386] Kohlberg sees morality as fairness and the application of moral rules:[387] that the concept of justice 'helps to concretize the concept of the moral by delimiting situations and attitudes to which criteria of the moral may be applicable'.[388] Fairness is closely linked to justice but while justice usually has been used with reference to a standard of rightness, fairness often has been used with regard to an ability to judge without reference to one's feelings or interests.[389] Conscience can be defined as a motivation to act morally which is partly based on an intuitive rejection of violation of 'fairness'.[390] Jerome sees the eagle as the 'spark of conscience' which he calls *synderesis*.[391]

---

[385] Immordino-Yang, M., Damasio, A. (2007), 'We Feel, Therefore We Learn: The Relevance of Affective and Social Neuroscience to Education,' *Mind, Brain, and Education*, 1 (1), p. 3.

[386] Ferguson, E., *et al.* (2014), 'Fast to forgive, slow to retaliate: intuitive responses in the ultimatum game depend on the degree of unfairness,' *PLOS ONE*, https://journals.plos.org/plosone/article?id=10.1371/journal.pone.0096344.

[387] Kohlberg L. (1969), 'Stage and sequence: The cognitive development approach to socialization,' in Goslin D. A. (ed.), *Handbook of Socialization Theory and Research*, New York: Rand Mc Nally, pp. 325–480.

[388] Forsyth D. R. (2020), *Making Moral Judgments: Psychological perspectives on morality, ethics and decision-making*, New York: Routledge, p. 50.

[389] Velasquez, M., et al. (2018), 'Justice and Fairness,' Markkula Center for Applied Ethics, at: https://www.scu.edu/ethics/ethics-resources/ethical-decision-making/justice-and-fairness/ Accessed 10/2/2020.

[390] Giubilini, Alberto (2016), 'Conscience,' *The Stanford Encyclopedia of Philosophy*, Winter 2016 Eedn, Edward N. Zalta (ed.), https://plato.stanford.edu/archives/win2016/entries/conscience/.

[391] St Jeromne (2017), *Commentary on Ezekiel*, Ancient Christian Writers series, New York: The Newman Press, p. 22.

A very fundamental balance in moral decision making and conscience seems to be that between an ethic of justice (or applying rules dispassionately) and an ethic of care (not hurting or harming and seeking mutually beneficial outcomes).[392] This mirrors the balance between authoritativeness and nurturance represented in the Living Creatures by the horizontal axis of the Lion and the Bull/Cattle (Figure 6). A recognised external authority tends to be obeyed even if it goes against conscience,[393] so it seems important in moral decision making that the exercise of authority is balanced against experiences of individuals affected.[394]

Rationality/Emotion Balance in Ezekiel

The effect of emotion on decision making was neglected in psychological models in the past, and it was thought that emotion interfered with cognitive processing. Recent research stresses that emotional input is essential for learning as well as important in decision making.[395] We have seen above that a principal function of the Creatures is to balance rationality and emotion and that they do this by subjection of their wills to that of the spirit. In Ezekiel, the rejection of God by the Israelites can be traced to a disconnection between rationality and emotion. This is played out in the oracle of Ez 36:16-38. A stressed motif in this oracle is that God's actions in human history are driven by revelatory aims: that his people and the world may know he is YHWH.[396] The Israelites

---

[392] Forsyth D. R. (2020), *Making Moral Judgments: Psychological perspectives on morality, ethics and decision-making*, New York: Routledge, p. 50.

[393] Milgram, S. (1963), 'Behavioral Study of Obedience,' *Journal of Abnormal and Social Psychology*, **67**, pp. 371–378; Milgram found that found that most people will obey a perceived external authority even if it goes against their conscience.

[394] Gilligan, C. (2014), 'Moral injury and ethic of care,' *Journal of Social Philosophy*, **45**, pp. 89–106.

[395] Zheng, Y. et al. (2017), 'The Influence of Emotion on Fairness-Related Decision Making: A Critical Review of Theories and Evidence,' *Front. Psychol.*, **8**, article 1592.

[396] Block, D.I. (1998), *The Book of Ezekiel Chapters 25-48*, Grand Rapids: Eerdmans, p. 366.

had however developed a flawed way of perceiving reality because their reason and emotions were not correctly integrated.[397] The Chariot and particularly the Creatures therefore may be an analogy of how, by allowing the spirit to correctly engage rationality and emotion, YHWH is able to reveal himself to humanity. This is further discussed in the chapter on Chapter 'The Interaction of Reason and Spirit'.

Insight and intuition in Making Quick Decisions

In terms of the cognitive axis in figure 6, it has been shown that insight and intuition,[398] (the character traits of the eagle), markedly aid decision making.[399] When a person arrives at an idea or a decision not by analytically inferring the solution, but by 'sensing' it, without being able to give reasons for it, this is 'intuition'; if the solution is realized all of a sudden without being able to report on the solution process, this is 'insight'. A major difference between rational decision making and insight and intuition are that the former is conscious, and the latter are subconscious.[400] Intuition helps an individual to make quick decisions, without engaging time-consuming rational analysis. The subconscious input into cognition from the 'eagle' could also include

[397] Launderville, D. F. (2007), *Spirit and Reason*, Waco: Baylor University Press, p. 348.

[398] Zander, T., *et al.* (2016), 'Intuition and insight: Two Processes that Build on Each Other or Fundamentally differ?,' *Front. Psychol.*, **7**, article 1395.

[399] Dane, E., *et al.* (2012), 'When should I trust my gut? Linking domain expertise to intuitive decision-making effectiveness,' *Organizational Behavior and Human Decision Processes*, **119** (2), p. 187.

[400] Zander, T., *et al.* (2016), *Ibid.*

supernatural spiritual insights.*[401] Strong emotions can influence intuition[402] and so it is important that intuition and emotion are separate, as they appear to be in the Living Creature character-forces: (in Figure 6 emotion represented by the bull/cattle face is on a different arm from intuition represented by the eagle).

Critique of Psychological Interpretation of Living Creatures

The criticism might be raised that applying a psychological interpretation to the different faces of the Living Creatures depends on an idiosyncratic interpretation of the four faces. In response to this, firstly, several commentators agree that the four faces on each of the Creatures, represents four countervailing forces.[403] These forces are (a) in a set relationship to each other; (b) depend on a specific anthropomorphizing of the animal concerned.

---

[401] In explaining differences in individual decision-making recent theorists have proposed a 'dual-process' theory, combining an evolutionarily primitive system of thought that relies on intuition and emotion (allowing quicker/ automatic responses) and a later developed slower system of reflection and analysis. Cf. Evans, J. (2008), 'Dual-processing accounts of reasoning, judgment, and social cognition,' *Annual Review of Psychology*, **59**, pp. 255–278. The cognitive axis of the eagle and human (Figure 6) fits in closely with this 'dual-process' theory. In addition, a recent 'fuzzy-trace' model of dual-process decision making incorporates reasoning with imprecise probabilities that is typical of fuzzy logic. Cf: Reyna, V. F., Brainerd, C. J. (2011), 'Dual Processes in Decision Making and Developmental Neuroscience: A Fuzzy-Trace Model,' *Dev Rev*, **31**, pp. 180–206. (Also see subsection 'Symbolic Reasoning Fuzzy sets and Fuzzy Logic' below). The 'fuzzy trace' model sees the two arms of the dual system as (a) verbatim representations that capture the precise, quantitative or surface form of situations (represented by the human face of the Creatures) and (b) gist representations that capture the bottom-line meaning of the situation which are vague and qualitative and based on anything that affects the essential meaning of a situation: e.g. education, culture, experience, emotion. This latter arm could capture spiritual insights or intuitions (represented by the eagle face).

[402] Bolte, A., *et al.* (2003), 'Emotion and Intuition: Effects of Positive and Negative Mood on Implicit Judgments of Semantic Coherence,' *J. Psychological Science*, **14**, pp. 416–421.

[403] Odell, M. S. (2005), *Ezekiel*, Smyth & Helwys Bible Commentary, Macon: Smyth & Helwys, p. 27

In terms of the key characteristics of the Creatures: rationality for the human; perspicacity for the eagle; authoritativeness for the lion and nurturance for cattle (See Appendix 12),[404] it is not necessary to prove that each trait is the principal or only anthropomorphised characteristic, but that it is widely recognisable as characteristic of the animal's behaviour. Our ability to make and understand metaphors appears to be an automatic process.[405] Servais says that animal mental qualities are not so much inferred as they are recognized, or directly perceived, by a human being who is engaged in a specific interaction.[406] For example, although 'lion' and 'eagle' are recognised as having a positive connotation when used metaphorically, most animal names used to describe human personality are uncomplimentary including cow or bull.[407] Since the character-forces chosen for the Living Creatures are widely attributed to these animals, and as detailed above, have been chosen as typical of them by several commentators, they do appear to be valid.

Living Creatures: Significance

In reply to the questions posed at the beginning of this chapter, the following conclusions about the Living Creatures can be made:

1. The Creatures are beings who appear to 'go between the divine and earthly realms'. They appear therefore to be part of both the spiritual and temporal worlds.

---

[404] It was noted above that the Hebrew word used for this animal has a meaning that encompasses bull, ox and cow.

[405] Hart, R. H. and Long, J. H. Jr (2011), 'Animal Metaphors and Metaphorizing Animals,' *Evolution: Education and Outreach*, 4, pp. 52–63.

[406] Servais, V. (2018), 'Anthropomorphism in Human–Animal Interactions: A Pragmatist View,' *Front. Psychol.*, 18, *https://doi.org/10.3389/fpsyg.2018.02590*

[407] Sommer, R., Sommer, B. A. (2011), 'Zoomorphy: Animal Metaphors for Human Personality,' *Anthrozoös*, 24, (3), pp. 237–24.

2. The use of composite creatures is not to be taken literally and the Creatures are not real beings such as angels, but several analogies can be discerned in them: (a) They have four major attributes of the Babylonian King: wisdom of man, fierceness of lion, fertility of the bull and swiftness and far-sightedness of the eagle. (b) They can represent four fundamental character-forces in humans:[408] rationality, authoritativeness (competitiveness/combativeness), nurturance and perspicacity (insight/intuition). (c) They can represent the four evangelists of the New Testament.*[409] (d) They are also four major forms of human pride that each of these animals portrays, but which through God, have been put under control : intellectual, power, possessions and spiritual.*[410] (e) They may collectively represent the priesthood. (f) They can represent four elements that are crucial for science (see 'Rationality, Science and Revelation'). (g) Finally, through their connection with the zodiac, the four Living Creatures represent the four corners of the cosmos representing the total control of YHWH over the universe.

---

[408] This thesis has not addressed the suggestion of Plato that the Living Creatures may represent different parts of the soul. Decision making appears to be situated in the heart but Aquinas attributes reason to the soul and says the emotion of love is felt as a passion by the soul: Gilson, E. (1956), *The Christian Philosophy of St Thomas Aquinas*, Indiana: University of Notre Dame Press, pp. 211, 273. The Creatures' character traits may therefore be embedded in the soul.

[409] Based on the psychological interpretation of the Creatures put forward in this thesis, the choice of which Evangelist to pair with each creature would be: Matthew – human (because his version is most rational, using frequent old testament sources as a foundation); Mark – lion (because his is the most authoritative as the earliest gospel and source for Matthew and Luke and the concise coverage in the limited size of his gospel); Luke – calf (nurturance in his treatment of birth of Christ in detail and his gentile-friendly approach); John – eagle (most spiritual). These are the same pairings as Jerome's and that the Church prefers, but apart from John, selected for different reasons.

[410] They can also be tentatively linked to the four fears of Ps 91:5-6: *terror of the night:* (loss of rational control – human); *arrow flying by day:* (power addiction – lion); *pestilence in the darkness:* (covetousness of worldly goods – bull); *plague at noon:* (spiritual abandonment – eagle). Cf. St Thomas More (1976), *A Dialogue of Comfort Against Tribulation*, Martz, L. L., Manley, F. (eds), New Haven: Yale University Press, p. xcvi.

3. The choice of the Creatures' faces and how they are presented, does reflect the *Sitz im Leben* of Ezekiel. The different animals represented by the four faces might also represent particular psychological/mental faculties in humans, through which the spirit acts to move man towards a moral goal. The four Living Creatures are therefore a model for key psychological forces involved with decision-making with conscious, and emotional as well as subconscious and spiritual inputs, and the way that these forces are arranged largely concurs with modern psychological dynamics. In addition, the Living Creatures may be an analogy for 'right reason', particularly superior reason.

Conclusion of the Chapter

This chapter argues that Babylonian pictography as reflected in four zodiac signs may have been an early source of the faces of the Living Creatures. The character forces imparted to the four faces have varied for Babylonian, Greek and New Testament Patristic commentators but over time certain core human characteristics stand out as a typical characterisation of each and can be labelled as follows: rationality (man), authoritativeness (lion), nurturance (cattle) and perspicacity (eagle). These character forces can be seen to function through two psychological axes: cognitive and emotional. The fixed position of the heads as the chariot moves forward is shown as a metaphor for decision-making; their straight movements, as a metaphor for right reason, through the interaction of reason with the spirit.

Chapter 4

Lower Tier: The Wheels of the Chariot

Overview of Chapter

This chapter reviews the Hebrew description of the wheels, their movements and the justification for including the wheels as a class of angels. The origins of the two Hebrew words for wheel used by Ezekiel, their etymological links and the different usage and meaning conveyed by them are analysed.

The wheels of the Chariot are probably the most 'iconic' part of the Chariot Vision as they give the vision the appearance of a 'chariot'. This chapter will give an analysis of the sections of Ezekiel dealing with the wheels including different interpretations of the 𝕸 Hebrew text in relation to the wheels and the difference between the two Hebrew words for wheel, *ôpan* (אוֹפַן) and *galgal* (גַּלְגַּל).

Movement of the Wheels

The wheels are described in Ez 1:17 as moving in any of the four directions 'when they went, they went to their four sides, they did not turn as they went' *'al-erba'at ribêhĕn bĕlektām yelēkû lô-yisăbbu bĕlektān* (עַל אַרְבַּעַת רִבְעֵיהֶן בְּלֶכְתָּם יֵלֵכוּ לֹא יִסַּבּוּ בְּלֶכְתָּן); whatever direction the front wheel faced, the others followed without veering as they moved (Ez 10:11). They have a spinning motion at times and in Ez 10:13 'the wheels were called whirling things':*[411] *laôpannîm lāhem qôra hagalgal* (לאופנימלהם קורא הגלגל) (Ez 10:13). Although the wheels do not appear to function as wheels, they do touch the ground. Ez 1:15, where the wheels are first mentioned, clearly describes a wheel on the ground: 'I noticed a wheel on the ground' *wa hinneh ôpan bā-areṣ* (וְהִנֵּה אוֹפַן אֶחָד בָּאָרֶץ).[412] The wheels were not attached with axles but were free-floating in character and they have no weight-bearing or directional function. Eichrodt says that the wheels do not seem to be necessary for the movement of the chariot and appear to be only a symbol for motion, since they mostly appear to hover over the ground.[413]

Symbolism of a Chariot

The concept of YHWH riding a chariot had already been enshrined in Psalm 104:3, ('He makes the clouds his chariot and rides on the wings of the wind').*[414] Boadt says the wheels indicate a war chariot, suggesting the presence of a divine warrior in battle. Wevers says that the kings of Israel and Judah rode in chariots as a sign of royal dignity, (1Ki 22:35; 2Ki 9:27; 2Ki 10:15) and throughout the Near East there was a widespread idea that God rode on a chariot.[415] The concept of YHWH riding a chariot may therefore be based on an anthropomorphic concept of kingship. Carley says that the wheels may symbolise a mobile platform on which the Ark of the Covenant could be moved

---

[411] *Hagalgal* הגלגל denotes a sense of movement rather than a static wheel, and is translated as 'rolling thing,' 'whirlwind,' or 'whirled dust', Cf. Strong, J. (2012), *Strong's Hebrew Dictionary of the Bible*, London: BN Publishing, p. 27. This is further discussed in a subsection below.

[412] Block, D. I. (1997), *The Book of Ezekiel Chapters 1–24*, Grand Rapids: Eerdmans, p. 98.

[413] Eichrodt, W. (1970), *Ezekiel*, London: SCM Press, p. 57.

[414] This psalm was thought to be written in 1015 BC towards the end of David's life, 423 years before Ezekiel wrote the first chapter of his book (592 BC).

[415] Wevers, J. W. (1969,) *Ezekiel*, London: Nelson, p. 47.

about (2Sa 6:3). A golden 'chariot' of the cherubim that spread their wings and covered the ark of the covenant is described in 1Ch 28:18.

Mystical Features of the Wheels

Described features of the wheels show that they are not normal chariot wheels*[416] but they do make a loud noise: 'the sound of a great rumbling' *waqôl rā'aš gādôl* (וקול רעש גדול) (Ez 3:13). In Ez 1:18 it says the wheels' rims were covered with eyes:[417] *wahaôpannîm melê'îm 'ênayim sābîb* (וְהָאוֹפַנִּים. מְלֵאִים עֵינַיִם סָבִיב) (Ez 1:18). According to one commentator this gives them a cosmic symbolism as they suggest stars.[418] Multiple eyes are well known from Assyrian statues of gods and reveal the all-seeing divine presence.[419] Brownlee says the eyes on the wheels show that the wheels were alive.[420] He says this is what led the Jews to list the wheels as a class of heavenly angels. (see subsection 'The Wheels as a Class of Angels'). The wheels are also twice described as like the gleaming of beryl (Ez 1:16; Ez 10:9); the four wheels having the same form (Ez 1:16); and they looked something like 'a wheel within a wheel' (Ez 1:16; Ez 10:10). This double wheel has been variously interpreted, (See Sections 2.7.3.;

---

[416] In the Jewish tradition, the four wheels of the chariot are said to be classes of angels: the *ôpannîm* and the *galgalim*. This interpretation appears to have emerged in Jewish Kabbalistic circles round the 10th century AD, Cf. Moore, G. F. (1927), *Judaism*, 3 vols, Cambridge, Mass.: Harvard University Press, p. 409; Scholem, G. (1961), *Major Trends in Jewish Mysticism*, New York: Schocken, p. 71; Recent commentators note that Ez 10, which ascribes apparent angelic characteristics to the wheels, is full of expansions and that at times these have become textually corrupt. See section 'The Wheels as a Class of Angels'.

[417] Sweeney, M. (2013), *Reading Ezekiel*, Macon: Smyth & Helwys, p. 65. says that the multiple eyes on the wheels could be interpreted as the eye of the divine presence looking throughout the world.

[418] Siegman, E. F. (1961), *The Book of Ezechiel*, New York: Paulist Press, p. 11.

[419] Boadt, L. (1990), 'Ezekiel,' *New Jerome Biblical Commentary*, London: Burns & Oates, p. 310.

[420] Brownlee, W. H. (1986), *Ezekiel 1–19*, World Biblical Commentary Volume 28, Waco: Word, p. 13.

2.7.4. also 'The Christocentric Interpretation'). Blenkinsopp says that the description of the wheels in Ez 1:15–21 is the most difficult and textually most obscure part of the chapter.[421] For example, the description that the rims of the wheels were 'tall and awesome' wĕgabbêhen wĕgōbah lāhem, wĕyir'â lāhem (וְגַבֹּתָם וְגֹבַהּ לָהֶם. וְיִרְאָה לָהֶם) in Ez 1:18 is not intelligible and may be a transcription error.[422] In Ez 10:12 the wheels are apparently given human and/or angel characteristics (flesh, backs, hands and wings). However, this could be due to a transcription error as discussed in the subsection below ('The Wheels as a Class of Angels').

The Spirit and the Wheels

The Spirit is totally in control of the wheels' movement and Brownlee[423] says that it is not axles that keep the wheels turning in proximity to the Living Creatures but the spiritual power of the Creatures themselves: 'When the Living Creatures moved, the wheels moved beside them and when the Living Creatures rose from the earth, the wheels rose' (Ez 1:19); 'Wherever the spirit would go, they would go and the wheels rose along with them' (Ez 1:20); the harmony of the Living Creatures extends to the wheels.[424] The phrase: 'for the spirit of the Living Creatures was in the wheels' ki rûaḥ hahayyâ bā ôpannîm (כי רוח החיה באופנים) is repeated in Ez 1:20 and Ez 10:17, emphasizing that normally inanimate objects appear to be alive due to the presence of the life-giving spirit

---

[421]   Blenkinsopp, J. (1990), *Ezekiel: Interpretation: A Bible Commentary for Teaching and Preaching*, Louisville: John Knox Press, p. 21.

[422]   Zimmerli says that this phrase 'their rims had height and fear' ('were tall and awesome' in NRSV) is a scribal error. Zimmerli, W. (1979), *Ezekiel 1*, Philadelphia: Fortress Press, p. 86.

[423]   Brownlee, W. H. (1986), *Ezekiel 1–19*, World Biblical Commentary Volume 28, Waco: Word, p. 13.

[424]   Odell, S. (2005) *Ezekiel*, Smyth & Helwys Bible Commentary, Macon: Smyth & Helwys, p. 33.

of YHWH.[425] Despite not being attached with axles, they kept pace with the Creatures and stayed beside them through every manoeuvre.

The Wheels as a Class of Angels

In the Jewish tradition, the four wheels of the chariot are said to be classes of angels: the *ôpannîm* and the *galgalim*.[426] This interpretation appears to have emerged in Jewish Kabbalistic circles round the 10th century AD.[427] Recent commentators however note that Ez 10, which ascribes apparent angelic characteristics to the wheels, is full of expansions and that at times these have become textually corrupt. A misapplication of the characteristics of the Living Creatures to the wheels may have occurred, particularly in Ez 10:14.[428] This is discussed in more detail after the text of Ez 10 relating to the wheels is given here (sections that have been questioned are in italics):

- '[8] The cherubim appeared to have the form of a human hand under their wings.

- [9] I looked, and there were four wheels beside the cherubim, one beside each cherub; and the appearance of the wheels was like gleaming beryl.

- [10] And as for their appearance, the four looked alike, something like a wheel within a wheel.

- [11] When they moved, they moved in any of the four directions without veering as they moved; but in whatever direction the front wheel faced, the others followed without veering as they moved.

- [12] *Their entire body, their rims, their spokes, their wings, and the wheels – the wheels of the four of them –were full of eyes all round.*

---

[425]   Block, D. I. (1989), 'The Prophet of the Spirit: the Use of RWH in the Book of Ezekiel,' *JETS*, **32**, pp. 27–49; p. 37.

[426]   Moore, G.F. (1927) *Judaism*, 3 vols, Cambridge, Mass.: Harvard University Press, p. 409.

[427]   Scholem, G. (1961), *Major Trends in Jewish Mysticism*, New York: Schocken, p. 71.

[428]   Halperin, D. J. (1976), 'The Exegetical Character of Ezekiel x 9–17,' *VT*, **26**, pp. 129–41; p. 137.

- [13] As for the wheels, they were called in my hearing 'the wheel-work' (*galgal*)

- [14] *Each one had four faces: the first face was that of the cherub, the second face was that of a human being, the third that of a lion, and the fourth that of an eagle.*

- [15] The cherubim rose up. These were the Living Creatures that I saw by the river Chebar.

- [16] When the cherubim moved, the wheels moved beside them; and when the cherubim lifted up their wings to rise up from the earth, the wheels at their side did not veer.

- [17] When they stopped, the others stopped, and when they rose up, the others rose up with them; for the spirit of the Living Creatures was in them.

- [20] These were the Living Creatures that I saw underneath the God of Israel by the river Chebar; and I knew that they were cherubim.

- [21] Each had four faces, each four wings and underneath their wings something like human hands.

Ez 10.14 states: 'Each one had four faces: the first face was that of the cherub, the second face was that of a human being, the third that of a lion, and the fourth that of an eagle.' The juxtaposition of this verse with the preceding one describing the wheels makes it seem that the wheels have these characteristics. Halperin thinks it was a later addition inserted to support the tendency to transform the wheels into a class of angels.[429] He says that the text in Ez 10:14 is likely an intrusion in 𝔐 which is a descriptive expansion that is lacking in 𝔊,[430] that appears to apply characteristics of the Living Creatures to the wheels.[431] The verse Greenberg says, in terms of the 'four faces' of the wheels mentioned in Ez 10:14, since this was not present in 𝔊, this is probably an error. This is

---

[429] The Midrash to Solomon's proverbs makes the *ôpannîm* and *galgalim* classes of angels. (*Midrash Mishle* (1884), f. 34a ff. ed. Buber, S., Cracow) quoted in Scholem, G. (1961), *Major Trends in Jewish Mysticism*, New York: Schocken, p. 71.

[430] Zimmerli, W. (1979), *Ezekiel 1*, Philadelphia: Fortress Press, p. 227.

[431] Wevers, J. W. (1969), *Ezekiel*, London: Nelson, p. 89 (footnote).

possibly due to speculation on the nature of the wheels in 𝔐.[432] He is here in agreement with Zimmerli. Sweeney says this passage is a reiteration of the appearance of the Living Creatures in Ez 1:10,[433] and does not apply to the wheels.

Ez 10:12 also appears to support the wheels as angels. The NRSV translation of Ez 10.12 ſtates: 'their entire body, their rims their spokes, their wings'. Greenberg's translation is: 'All their flesh and their hands and their backs and their wings' *wekāl beśārām wĕgabbêhem wîdêhem wekanpêhem* (וְכָל־בְּשָׂרָם, וְגַבֵּהֶם, וִידֵיהֶם,וְכַנְפֵיהֶם).[434] Some authors ascribe these characteriſtics to the wheels in a 'myſtic' sense.[435] but others ascribe them to the 'cherubim'.[436] Wevers, however, says that Ez 10:8-17 is a series of expansions based on Ez 1:18 that is added to ensure the identity of the cherubim of Ez 10:8 with the Living Creatures of the initial vision in Ez 1:5. Some of these expansions have become corrupt and the result is peculiar or even absurd.[437] Zimmerli points out that the phrase 'all their flesh' (we*kāl beśārām*) is missing in 𝕲.[438] He feels that there has been a secondary elaboration of Ez 10:12 to include characteriſtics of the cherubim. For example, the description that appears to relate to the cherubim in Ez 10:12 is similar to the description in Ez 1:18 that describes the wheels. In Ez 10:12 the wheels are described as 'having eyes all around' *melê'îm 'ênayim sābîb larba'tām* (מְלֵאִים עֵינַיִם סָבִיב. לְאַרְבַּעְתָּם) which appears to be a repeat of Ez 1:18 *melê'ōt ênayim sabîb le'arba'tān* (מְלֵאֹת עֵינַיִם סָבִיב--לְאַרְבַּעְתָּן). However,

432   Greenberg, M. (1983) *Ezekiel 1-20*, New York: Doubleday, p. 47.

433   Sweeney, M. A. (2013), *Reading Ezekiel*, Macon: Smyth & Helwys, p. 65.

434   Greenberg, M. (1983), *Ibid.,* p.179.

435   Greenberg, M. (1983), *Ezekiel 1-20*, New York: Doubleday, p. 182.

436   Greenberg, M. (1983), *Ibid.,* p. 179.

437   Wevers, J. W. (1969), *Ezekiel*, London: Nelson, pp. 88-89 (footnote).

438   Ezekiel 10:12 from 𝕲: '[missing phrase] καὶ οἱ νῶτοι αὐτῶν καὶ αἱ χεῖρες αὐτῶν καὶ αἱ πτέρυγες αὐτῶν καὶ οἱ τροχοὶ πλήρεις ὀφθαλμῶν κυκλόθεν τοῖς τέσσαρσιν τροχοῖς αὐτῶν' (and their backs and their hands and their wings...) Zimmerli, W. (1979), *Ibid.*, p. 227.

an elaboration of the word *gōbah: wĕgōbah lāhem* (וְגֹבַהּ לָהֶם),[439] 'and they were tall' from Ez 1:18 becomes *wĕgabbêhem* (וְגַבֵּהֶם) 'their backs' in verse Ez 10.12, referring to the cherubim. Greenberg says that the use of the word 'flesh' to describe cherubim is however surprising.[440] Carley says that there has been a major expansion of Ez 10:9-12 and later insertion of verses Ez 10:13-15. He says that the intrusions are meant to try and harmonise the various features of the wheels in the first vision in chapter 1 and that Ez 10:20-21 appears to be an editorial comment that clarifies Ez 10:12 by ascribing the faces, hands and wings to the cherubim rather than to the wheels.[441]

In agreement with the above, in this thesis, verse 12 is taken to be corrupted and therefore the description is not used. Verse 14 is taken to apply to the Living Creatures and not to the wheels. Calling the Living Creatures 'cherubim' in chapter 10 could have been part of the tendency to make the moving components of the Chariot into angels. As explained above the Living Creatures have rationality and therefore probably should not be classed as true angels. The description of the wheels in Ez 1:18 as 'tall and awesome' is taken to be a transcription error.

Ôpan and Galgal: Significance of the Hebrew Words for 'Wheel'

The Hebrew words *ôpan* (אוֹפָן) and *galgal* (גַּלְגַּל) are both found in the Chariot of Yahweh vision and often both translated in the English Old Testament versions as 'wheel'. Their significance in the Hebrew language appears to be different however:

---

439 Zimmerli feels that this first part of Ez 1:18 is badly disturbed and *wĕgōbah lākēm* should be *wĕgābôt lākēm*. Zimmerli, W. (1979), *Ibid.*, p. 227.

440 Greenberg, M. (1983), *Ibid.*, p. 182.

441 Block, D. I.(1997), *The Book of Ezekiel chapters 1-24*, Grand Rapids: Eerdmans, p. 324.

Galgal (גַּלְגַּל)

The hebrew word *galgal* originates from the verb *galal* (to roll), by redupli-cation.[442] and primarily means 'wheel'. *Galgal* also means wheel in Chaldean[443] and in Aramaic. The origin of *galgal* from the verb 'to roll' (*galal*) is important in that *galgal* denotes a sense of movement rather than a static wheel, and is translated as 'rolling thing', 'whirlwind' or whirled dust'.[444] *Galgal* in Ps 77:18 has been translated as 'in heaven' in the King James Bible. However, other Bible versions put the translation as 'rumbled in the sky' (God's Word Translation) or 'was all around' (Jubilee Bible), again suggesting a circular motion component to the word *galgal*. *Galgal* is often used for the wheels of a war chariot (Is 5:28; Je 47:3)

Brown-Driver-Briggs[445] gives 'whirlwind' (psalm 77:18), 'whirling dust' (Ps 83:13 and Is 17:3) as translations for *galgal*. 'Wheelwork' is also used in Ezekiel (Ez,10:6; Ez 10:13) (Jewish Publication Society Tanakh 1917; Christian Standard Bible) but other translations (English Standard Version) use 'whirling wheels' rather than 'wheelwork' in Ezekiel. Zimmerli, states that there is no doubt that *galgal* here signifies 'wheels'.[446] *Galgal* also means a whirlwind in Syriac.

Galgal, can also stand for the straw, chaff or husk that is driven by a whirl-wind,[447] as in Is 17:13[448]. Interestingly, the Hebrew word for skull, (גֻּלְגֹּלֶת) *gulgôlet*, is derived, like *galgal*, by reduplication from *galal* (to roll). This ap-pears to be because a skull is round. *Gulgôlta*, the Aramaic name for the 'Place

---

[442]   Strong, J. (2012), *Strong's Hebrew Dictionary of the Bible*, London: BN Publishing, p. 27.

[443]   Strong, J. (2012) *Ibid.*, Concordance, 1535.

[444]   Strong, J. (2012) *Ibid.*, p. 27.

[445]   Brown, F., Driver, S., Briggs, C. (1991), *Brown-Driver-Briggs Hebrew and English Lexicon*, Oxford: Clarendon Press.

[446]   Zimmerli, W. (1979), *Ezekiel 1*, Philadelphia: Fortress Press, p. 251.

[447]   Gesenius (1857), *Hebrew–Chaldee Lexicon*, London: Bagster

[448]   In Ps 83 (83:14) *galgal* is used for whirling dust: 'O my God make them like whirling dust', *'ĕlōhây, šîtēmo kācagalgal* (אֱלֹהַי. שִׁיתֵמוֹ כַגַּלְגַּל). Also in Isaiah 17:13 'and like whirling dust before the storm' *Wĕkagalgal lipnê sûpah* (וּכְגַלְגַּל. לִפְנֵי סוּפָה).

of the Skull', is a derivation from *gulgôlet*. Golgotha, the Biblical name for the place where Christ was crucified is a Greek corruption of *Gulgôlta*: Γολγοθᾶ (Golgotha). Another derivation of *galal* is Galilee (Γαλιλαία). Strong's Concordance shows that Galilee derives from *gĕlilah* (גְּלִילָה) which means a circuit, or a territory, from whence Galilee. *Galal* gives rise to *gāl*, which means a wave or a heap, but can mean a ruin as in Is 25:2: 'for you have made of a city a ruin' *kî śamta mê'îr lagāl* (כִּי שַׂמְתָּ מֵעִיר לַגָּל).

4.6.2. Ôpan (אוֹפָן)

According to Strong, the Hebrew word *ôpan* originates from an unused root meaning to revolve.[449] Brown Driver Briggs suggests a derivation from *pana*, (to turn) but the root *ôpp*, (to surround or encompass) is more likely.[450] *Ôpannîm* is the usual Hebrew word for wheels, but in later Jewish tradition, a choir of angels was called *ôpannîm*.[451] This may have its origin in Ezekiel, due to a misinterpretation of Ez 10:14 (See 'Wheels as a Class of Angels'). *Ôpannîm* is used to describe the wheels of the Chariot of YHWH in Ezekiel, where it is used 21 times (as opposed to *galgal* only used 3 times). *Ôpannîm* is used for chariot wheels, when it is talking about their construction rather than their motion, for example, Ex 14:25. It is also used to describe the wheels of the ten bases beneath the lavers in Solomon's temple. A derivative word from *ôpan*, *ôpen* (אֹפֶן) is translated as: a turn, a right circumstance or condition. It is a hapax in the Old Testament.[452]

---

[449] Strong, J. (2012), *Strong's Hebrew Dictionary of the Bible,* London: BN Publishing, p. 9.

[450] Block, D. I. (1997), *The Book of Ezekiel Chapters 1–24,* Grand Rapids: Eerdmans, p. 98. (footnote)

[451] Siegman, E. F. (1961), *The Book of Ezechiel,* New York: Paulist Press, p. 11.

[452] Proverbs 25:11 '*dābār dābur 'al-āpenāw*' דָּבָר דָּבֻר עַל-אָפְנָיו (like a word spoken in the right setting.)

Discussion

Ôpan is used much more frequently in the Old Testament than galgal (36 times versus 11 times). Unlike galgal which may have different meanings (whirling, stubble etc., see above), ôpan is almost invariably translated as wheel. In Ezekiel galgal is only used three times to describe a wheel. Two of these three uses of galgal are in association with the taking of burning coals from between the wheels. Interestingly in Daniel (Da 7:9), galgal is used to describe the wheels of the throne of the 'ancient of days' that were of burning fire.[453]

Greenberg says that it is impossible to establish a semantic distinction between the words ôpannîm and galgal. Both are equally appropriate for the divine vehicle. Galgal is used as a collective for chariot wheels. In contradistinction to galgal, ôpan has a very concrete connotation, Ôpan is also used for chariot wheels and for the wheels of the laver stand in the temple. The word is used for the wheels of the lavers in Solomon's temple, items which were very carefully measured (1Kg 7:38). The word ôpannîm is used in the initial Chariot Vision description (Ez 1:15) but Ezekiel hears the voice of a person (?YHWH) mention the word galgal: Ez 10.13 says: 'I overheard the wheels' (ôpannîm) being identified as "galgal",' and this is how it comes to be used.[454] Block interprets galgal as 'casters' a more specific term which has a collective sense. Zimmerli wonders whether galgal may have a special sense: possibly wheels which were separated from the Ark of the Covenant after 597 BC (Je 3:16)[455] or an incense brazier or a war chariot (Is 5:28). He says that it may refer to a visible object seen with wheels. He says this cannot be answered with certainty and it remains speculative. Zimmerli gives an alternative interpretation for

---

[453]   Daniel 7:9 'galgîlôhi nōr dalîq' גַּלְגִּלּוֹהִי נוּר דָּלִק (its wheels were burning fire) Daniel 7:9 'I kept looking Until thrones were set up, And the Ancient of Days took His seat; His vesture was like white snow And the hair of His head like pure wool His throne was ablaze with flames, Its wheels were a burning fire.'

[454]   Ezekiel is 'directed' by the voice, (possibly YHWH) to use 'galgal' instead of ôpannîm stressing that the use of galgal is unusual and likely specific for the situation.

[455]   Zimmerli, W. (1979), Ezekiel 1, Philadelphia: Fortress Press, p. 251.

galgal as an 'incense brazier'. He states that the interpretation of *galgal* gives 'considerable impulse to the elaboration of the wheel speculation'.[456] He does not elaborate on this speculation however.

In summary, of the two Hebrew words for wheel, *ôpan* and *galgal*, *galgal* has connotations of rapid whirling movement. The way *galgal* is used in Ezekiel suggests that it is capturing a characteristic of the wheels of the Chariot of YHWH that is distinct from a movable structural support for the Chariot. The fact that it is used in relation to a fire both in Ezekiel and in Daniel suggests a spiritual or transcendent character of the wheels.

Two characteristics of the wheels that appear to be of central importance in the context of this thesis are that firstly, the wheels are the only part of the vision that incorporate artificial construction.[457] Secondly, they are critically situated at the junction between the divine and earthly realms[458] as the wheels contact the ground, (Ez 10:16). The wheels are driven by the spirit, as the Living Creatures are guided by the spirit and 'the spirit of the Living Creatures was in the wheels' (Ez 1:20). The wheels may therefore link the spirit to measurable finite distances that the Chariot travels over the earth in its 'to-and-fro' (Ez 1:14) movements. Since measurement is a rational way of gaining knowledge, in contrast with the 'superior reason' of the Living Creatures (see section 3:15), the wheels' movements might reflect 'inferior reason'.[459] Since this concrete or measurable revelatory knowledge is initiated by the action of the spirit, it is in accordance with 'divine science'. The link between measurement and rationality in Ezekiel is discussed further in 'Rationality and Measurement in Ezekiel'.

---

[456] Zimmerli, W. (1979), *Ibid.*, p. 251.

[457] Eichrodt, W. (1970), *Ezekiel*, London: SCM Press, p. 57

[458] Odell, S. (2005), *Ezekiel*, Smyth & Helwys Bible Commentary, Macon: Smyth & Helwys, p. 29.

[459] Pullicino, P. (2021), 'For the spirit of the living creatures was in the wheels: Subjection of the Wheels to the Spirit of Rationality, (Ezekiel 1:20),' *Polish Journal of Biblical Research*, **19–20**, pp. 91–104.

Conclusion of the chapter

The wheels are responsible for the vision being called a chariot. The Hebrew word *galgal* has connotations of rapid whirling movement which suggests the wheels have a transcendent character. The wheels incorporate artificial construction and contact the ground and move measurable distances and may be an analogy for inferior reason and of 'divine science'.

Chapter 5

Upper Tier: The Glory of the God of Israel

Overview of chapter

This chapter reviews the Hebrew text relating to the 'likeness of a man' image. In particular the origin and meaning of the word *ḥašmal*. The significance of the fire above and below the loins, the rainbow and the crystal firmament are also examined.

The 'likeness of a man' Image

The upper tier of the Chariot Vision, which includes the 'Glory of the God of Israel'*[460] *kābôd 'ĕlōhê yiśrā'ēl* (כְּבוֹד אֱלֹהֵי־יִשְׂרָאֵל) is delimited by the words 'above' *'al* (עַל), (Ez 10:1), or *milmā'lah* (מִלְמָעְלָה) (Ez 10:19), which describe its position in relation to the Living Creatures.[461] The upper tier includes the

---

[460] Ezekiel appears to keep the designation 'Glory of the God of Israel' for the appearance above the Living Creatures (Ez 9:3; 10:19), whereas he uses 'Glory of YHWH' for the whole Chariot structure (Ez 1:28; 11:23).

[461] Cook, S. L. (2004) 'Cosmos *Kabod* and Cherub: Ontological and Epistemological Hierarchy in Ezekiel,' in *Ezekiel's Hierarchical World: Wrestling with a Tiered Reality*, C. L. Patton and S. L. Cook (eds), Atlanta: Society of Biblical Literature, 179–198; p. 185.

sapphire throne (Ez 10:1), 'a form like the appearance of a human being'*[462] *děmût kěmar'ēh ādām* (דְּמוּת כְּמַרְאֵה אָדָם), the crystal firmament (Ez 1:22) and the rainbow (Ez 1:28). The enthroned likeness of a man is the crowning focus of the Chariot Vision which elicits the Chariot's appellation of the 'Glory of YHWH', *kābôd YHWH* (כְּבוֹד יְהוָה) (Ez 1:28).

Questions that arise about this figure are: What is the identity of the figure seated on the throne? What is the meaning of the amber *ḥašmal* and its enclosing fire (Ez 1:27)? What is the connection between this figure and the commands made to Ezekiel? What is the meaning of the fire below the loins? What is the connection between the figure and the spirit and what is the revelatory significance of the figure (for this last question see 'Spirit in Ezekiel and the Chariot')?

A central question, therefore, is the identity of the 'likeness of a man'. We have discussed that the Chariot Vision is likely an objective construct that reflects real attributes of God, or denotes the presence of God - but does this figure represent God? The figure was described as having a gleaming amber appearance from the loins upwards and the appearance of fire, downwards from the loins (Ez 1:27). The appearance of the Chariot causes the prophet to fall on his face.[463] The inference is that Ezekiel believes YHWH to be present in front of him.[464] Immediately after this Ezekiel says in Ez 2:1 'and he said unto me'. On four separate occasions (Ez 2:1; Ez 3:24; Ez 8:5; Ez 43:4), the appearance of the 'Glory of YHWH' vision is followed by a series of commands or statements by a voice, that Ezekiel takes to be the voice of God, and he

---

[462] *Děmût kěmar'ēh ādām* is most faithfully translated as 'a form like the appearance of a human being': Block, D. I. (1997), *The Book of Ezekiel Chapters 1–24*, Grand Rapids: Eerdmans, p. 102. In this thesis the phrase 'likeness of a man' is used instead to simplify the text.

[463] The prophet falls on his face three times: Ez 1:28; Ez 3:23 and Ez 43:3. Each occurrence is at the appearance of the Glory of the God of Israel.

[464] Block, D. I. (1997), *The Book of Ezekiel Chapters 1–24*, Grand Rapids: Eerdmans, p. 105.

denotes the speaker as the Lord YHWH: *'ădōnay YHWH* (Ez 2: 4).*[465] It is not specified however that the voice is that of the likeness of a man at the top of the Chariot Vision.

In Ez 8:2, the prophet sees the likeness of a man with gleaming amber (*ḥašmal*) above the loins and fire below, presumably the same figure that he had seen initially on the throne in Ez 1:27. He appears to Ezekiel in his house in Babylonia, but without the remainder of the Chariot Vision. The prophet is then taken to Jerusalem in a vision and, 'the Glory of the God of Israel' was there, 'according to the vision that I saw in the plain'. In a continuation of this vision, (of Ez 8:2) which becomes the Vision of the Executioners, in Ez 9:3, the prophet hears a voice, which he takes to be that of God,[466] talking to the executioners. Ezekiel then falls on his face and says, 'Ah Lord YHWH'. It is reasonable to assume that the person talking to Ezekiel was the figure that was previously seen on the Chariot Vision (Ez 8:2) since the words are spoken in a continuation of this vision.*[467] In support of this, in Ez 10:1 the throne at the top of the Chariot Vision is described without the mention of a figure. However, when the Chariot of YHWH leaves Jerusalem at the end of this vision, the 'Glory of the God of Israel' is again seen on it (Ez 10:19).

In summary, the seated figure on the top of the Chariot of YHWH, is called by Ezekiel the 'Glory of the God of Israel' and appears to represent the majesty or glory of the Lord.*[468] Although Ezekiel never explicitly calls this figure

---

[465] Ezekiel uses 'the Lord YHWH' 217 times, particularly at the beginning or conclusion of divine speeches and Block says it is used by Ezekiel to show both God's personal name and official title and establishes Ezekiel's authority: Block, D. I. (1997), *The Book of Ezekiel Chapters 1–24*, Grand Rapids: Eerdmans, p. 30.

[466] Ez 9:1 'Then he cried in my hearing...' also Ez 9:3 'the glory of the God of Israel was gone up from the cherub,... and He called to the man clothed in linen...'

[467] The text does not however explicitly say that the man with the gleaming amber above his loins and fire below was the person speaking, or that this man was present at that particular time.

[468] Other representations of God include that of Isaiah who saw a vision of the Lord sitting on a throne (Is 6:1) and Daniel had a vision of 'one like a son of man' being presented

'Lord' or 'God' he does identify God as the person who talks to him following the appearance of the Chariot Vision.

The 'likeness of a man' image is described in detail on two occasions. Firstly, in Ez 1:26 and again in Ez 8:2. These descriptions differ slightly so the details are compared here:

Ez 1:26	Ez 8:2
Figure like the appearance of a human being	A figure having the appearance of a man
From the appearance of his loins upward	And from his loins up
I saw the like of *ḥašmal*	Like *ḥašmal*
Having something of the appearance of fire	Something with a brilliant (זֹהַר *zōḥar*)
enclosing it *kĕmar'ēh eš bêt lah sābîb*	appearance
† (כְּמַרְאֵה־אֵשׁ בֵּית־לָהּ סָבִיב)	
From the appearance of his loins downward	From the appearance of his loins down
I saw something with the appearance of fire	was fire
wenōgah (וְנֹגַהּ)	
Like the appearance of a bow	

† Zimmerli states that the phrase 'having something of the appearance of fire enclosing it' is not present in ⑥ and that it must be removed as it introduces the appearance of fire to the upper part of the body, but it should only belong to the lower body. Zimmerli, W. (1979), *Ezekiel I*, Philadelphia: Fortress Press, p. 88.

As discussed above, the Chariot Vision can be interpreted as an analogy of the way in which God reveals himself to the world. The likeness of a man on top of the Vision, is most likely therefore an analogy rather than a direct portrayal of God. Fisch states that a divided likeness (the figure on the Chariot of YHWH is divided at the loins) cannot represent God so that the figure is unlikely to be a direct vision of God himself. However, as discussed above, there is a suggestion that the figure may be an analogy of God and may denote his presence. In addition, the whole Chariot of YHWH Vision, (including the

before 'One of Ancient of Days' (Da 7:13). In Revelation there are two descriptions: of one 'seated on a throne' (Re 4:2) and one 'seated on the cloud, one like a son of man'. (Re 14:14) The exact identity of these different divine representations is however unclear.

Living Creatures and wheels), appears to denote the presence of God and his Glory. Ezekiel's action of 'falling on his face', implies he recognises the spiritual presence of the Lord in this vision. The question of whether the figure could be an analogy of Christ is discussed below.[469]

Hashmal (ḥašmal)

The word ḥašmal is mentioned three times in Ezekiel - in connection with the Chariot Vision. The significance of ḥašmal has been pondered over and has deep mystical significance in Jewish Kabbalistic interpretations.[470] The Kabbalistic interpretation*[471] is that the ḥašmal represents a pivotal portal of the divine presence into the Chariot. Ḥašmal is a hapax legomenon possibly relating to the Akkadian word elmešu, which was a brilliant precious stone.[472] An etymological link has been suggested between the Hebrew word ḥašmal and the Elamite ismalu, Akkadian esmaru, Ugaritic azmar and Old Persian ismaruv: these words designated a kind of metal, probably bronze.[473]

---

[469]   Fisch, S. (1950), *Ezekiel*, London: The Soncino Press, p. 8.

[470]   *Talmud Mas. Chagigah*, 13a 49,50: The Rabbis taught: There was once a child who was reading at his teacher's house the Book of Ezekiel, and he apprehended what *Hashmal* was, whereupon a fire went forth from *Hashmal* and consumed him.

[471]   In the literature of the mediaeval Kabbalah, the Hashmal belongs to the 'Yetsiratic' world. According to a modern Bible commentator (the celebrated Russian Hebraist, M. L. Malbim, 1809–1879) the word signifies 'the *Hayot* [i.e. 'Living Creatures' of Ezekiel] which are the abode [or camp] of the *Shechinah* [i.e. Divine Presence] where there is the 'still small voice.' It is they who receive the Divine effluence from above and disseminate it to the movers of the 'wheels' of Ezekiel's Chariot. Abelson, J. (2001), *Jewish Mysticism: An Introduction to the Kabbala*, reprinted from 1913 version, London: Global Grey, p. 38.

[472]   Garfinkel, S. P. (1983), *Studies in Akkadian influences in the Book of Ezekiel*, Ann Arbor: University Microfilms, p. 81.

[473]   Montgomery, J. A. (1936) *apud* Kent, R. G. 'The Present State of Old Persian Studies,' *JAOS* **56**, pp. 208–225.

Bodi reviews several Akkadian and Elamite potential origins for *ḥašmal*.[474] Zimmerli[475] states that *ḥašmal* translates as *electrum*, a bright mixture of gold and silver.[476] He says that the interpretations relating *ḥašmal* to bronze are probably incorrect, since Driver[477] based his comment on 𝔐 and this is not supported in 𝔊, and there is nothing in that text to suggest metal in a molten state.

Ḥašmal is first mentioned early (Ez 1:4) in the inaugural vision which introduces a flashing fire with an amber colour at its centre 'the colour of electrum out of the centre of the fire' *haḥašmal mittok haeš* (כְּעֵין הַחַשְׁמַל. מִתּוֹךְ הָאֵשׁ). This initial glimpse of the *ḥašmal* is like a preparatory announcement or preview of the 'Glory of the God of Israel' (Ez 10:19), that is described in more detail later (Ez 1:26-28). *Ḥašmal* is again mentioned in connection with the 'likeness of a man' description in Ez 8:2. The description of the appearance above the waist is given as 'the appearance of brightness, the colour of electrum (*ḥašmal*).' *kĕmarē zōhar kĕ'ēn haḥašmalah* (כְּמַרְאֵה-זֹהַר כְּעֵין הַחַשְׁמָלָה). *Ḥašmal* is particularly connected with the upper part of the body of the man in the first (Ez 1:27) and second (Ez 8:2) descriptions. Jerome sees the region above the loins as the region 'where the senses and reason abide' and says that therefore this area does not need fire but just 'precious and pure metal'. By this Jerome likely means the heart, as the heart is the traditional location of thought and feel-

---

474   Bodi, D. (1991), *The Book of Ezekiel and the Poem of Erra*, Orbis Biblicus et Orientalis, no. 104, Göttingen: Vandenhoeck & Ruprecht, p. 83.

475   Zimmerli, W. (1979), *Ezekiel 1*, Philadelphia: Fortress Press, p. 122; citing Koehler-Baumgartner from Lucas, A. (1926), *Ancient Egyptian Materials*, New York: Longmans, pp. 84–86.

476   Electrum is a naturally occurring alloy of silver and gold with an amber colour which can be made artificially.

477   Driver, G. R. (1951), 'Ezekiel's Inaugural Vision,' *VT*, **1**, pp. 60–62; p. 60.

ings.[478] *Ḥašmal* therefore could represent the heart. Almoſt all early Chriſtian commentators see the *ḥašmal* as connected in some way with Chriſt.[479]

Fire in the 'Likeness of a Man' Figure

There are two separate areas described as 'fire' in the 'likeness of a man' figure. The dividing line for these two areas is 'the appearance of his loins'[480] *mimmar'ēh motnāyw* (מִמַּרְאֵה מָתְנָיו) Ez 1:29. Above the loins was the amber appearance of *ḥašmal*, surrounded by the appearance of fire.[481] In addition, in Ez 8:2, the *ḥašmal*, is given 'the appearance of brightness' (*zōhar*). Below the loins was 'the appearance of fire' *kĕmar'ēh ēš* (כְּמַרְאֵה־אֵשׁ) in Ez 1:29 and 'fire' in Ez 8:2. Brightness (*nōgah*), is also described as being around the figure of the man in Ez 1:27. Below the loins, the fire is a ſtronger, fiercer image; above the loins, the appearance is comparatively gentler. Eichrodt calls the appearance above the loins 'a calm gleam of heavenly brightness' whereas below the loins it is 'a menacing might of licking flames'.[482]

In Ez 1:27, the *ḥašmal* is seen in the upper half of the likeness of a man on a throne, surrounded by fire: 'as the colour of electrum as the appearance of

---

[478] St Jerome (2017), *Commentary on Ezekiel,* Ancient Chriſtian Writers, New York: The Newman Press, p. 33; In Hebrew, there is no separate location for noetic activity and the heart is the place in which noetic and affective activities converge: Launderville, D. F. (2007), *Spirit and Reason,* Waco: Baylor University Press, p. 54.

[479] Chriſtman, A. R. (2005), *'What did Ezekiel See?,'* Leiden: Brill, p. 30: Eusebius says that the Logos 'is compared to the electrum (*ḥašmal)* which is more precious than anything else.' Gregory the Great, (*Hom Ez,* I.8.25: 4–11) sees the *ḥašmal* as prefiguring the incarnation. *Ḥašmal* can be equated with agapē, (*ἀγάπη*) as the love of God for man, which particularly in Chriſt, is centred on the heart.

[480] *Motnāyw* (מָתְנָיו) means 'loins' or 'waiſt' according to the *New American Standard Hebrew–Aramaic Exhaustive Concordance of the Bible* (1997).

[481] Zimmerli says that the phrase 'including the surrounding fire', was a later addition, not present in 𝔊. There should therefore probably not be any reference in the original text, to fire as describing the upper body. Zimmerli, W. (1979) *Ezekiel 1*, Philadelphia: Fortress Press, p. 88.

[482] Eichrodt, W. (1970), *Ezekiel,* London: SCM Press, p. 58.

fire round about enclosing it'. *w'erē kĕ'ên ḥašmal kĕmar'ēh eš bêt lah sābîb* (וָאֵרֶא כְּעֵין הַשְׁמַל, כְּמַרְאֵה-אֵשׁ בֵּית-לָהּ סָבִיב). Zimmerli feels that the phrase 'the appearance of fire round about and enclosing it' is a later addition as it is not present in 𝔊. When *ḥašmal* is again mentioned in Ez 8:2, there is no surrounding fire. If we accept Zimmerli's[483] and Allen's[484] objections to fire being used as a description of the upper part of the body of the likeness of a man vision, then the body of the man has the appearance of *ḥašmal* above the waist and fire below the waist. However, when *ḥašmal* was first seen in the storm cloud (Ez 1:4), at the beginning of the Chariot Vision, it appeared 'out of the midst of the fire'.[485] This is against the views of Zimmerli and Allen that the fire surrounding the *ḥašmal* in 1:27 was a later addition. In this thesis the fire surrounding the *ḥašmal* is not treated as an addition.

The question is, what does this 'fire' in connection with the 'likeness of a man' signify? Jerome sees the region below the loins as the region where intercourse and procreation occur, where 'the incentives require the purification of flames.'[486] The fire below the loins could therefore represent sexuality.[487] The fire surrounding the *ḥašmal* appears to be illustrating or accentuating the *ḥašmal* and it is discussed further under 'The Spirit in Ezekiel and the Chariot'.

---

[483] Zimmerli, W. (1979), *Ibid.,* p. 88.

[484] Allen, L. C. (1994), *Ezekiel 1–19*, World Biblical Commentary 28, Dallas: Word, p. 9.

[485] Zimmerli takes this phrase also as an addition. Zimmerli, W. (1979), *Ezekiel 1*, Philadelphia: Fortress Press, p. 88.

[486] St Jerome (2017) *Commentary on Ezekiel*, Ancient Christian Writers, New York: The Newman Press, p. 33; Eusebius also equates the fire below the loins with sexuality: Christman, A. R. (2005), *'What did Ezekiel See?,'* Leiden: Brill,. p. 30 (footnote).

[487] Eusebius sees the fire below the loins symbolizing sexuality. A link between intimate love, eros (ἔρως) and fire is found in the Song of Songs (So 8:6): 'For love is strong as death... the flashes thereof are flashes of fire, a very flame of the Lord', (רְשָׁפֶיהָ--רִשְׁפֵּי. אֵשׁ שַׁלְהֶבֶתְיָה) *rešapēhā rišpê eš šalhebetyāh*).

The Rainbow

The brightness that was around the son of man figure in Ez 1:28 is described as 'like the appearance of a bow that is in the cloud in the day of rain' *kĕmar'ēh haqešet ašer yhyĕh bĕ'ānan bĕyom hagēšem* (כְּמַרְאֵה הַקֶּשֶׁת אֲשֶׁר יִהְיֶה בֶעָנָן בְּיוֹם הַגֶּשֶׁם). In other parts of the Bible, apart from the well-known occurrence in the story of Noah (Ge 9:13), a rainbow that looked like an emerald is seen round the one seated on the throne in Re 4:6. A rainbow is present over the angel in Re 10:1 who comes wrapped in cloud at the end of the world with a scroll for John to eat. The rainbow is a divine theophanic sign and the rich colour is evidence of the visionary nature of the experience.[488]

Firmament of Crystal

The appearance of the Glory of God of Israel as the 'likeness of a man', recalls Christ. The figure was seated on a throne which was on a 'firmament'. The appearance of the throne was like sapphire. Blue was the colour of Aaron's robe and is one of the three colours of blue, purple and scarlet of the priestly ephod. (Ex 39:1). Blue is therefore a holy colour in the Old Testament and the sapphire throne could be a representation of the presence of YHWH. In the Old Testament, the 'God of Israel' appears on a pavement of sapphire stone (Ex 24:10), which has similarities to the appearance of the likeness of a man on the sapphire 'throne' in the Chariot of YHWH vision.[489] The firmament above the chariot is seen as a copy of the firmament in Genesis (Ge 6:1) that was set up to divide the earthly waters from heavenly waters above which YHWH is enthroned.[490] In Ez 1:22 the firmament is described as being the colour of terrible ice: *rāqî'a kĕ'ên haqqerah hannora'* (רָקִיעַ כְּעֵין הַקֶּרַח הַנּוֹרָא). However,

---

[488] Joyce, P. M. (2007), *Ezekiel: A Commentary*, New York: T & T Clark, p. 74.

[489] This could be why Ezekiel reserves the appellation 'glory of the God of Israel' for the enthroned figure.

[490] Eichrodt, W. (1970), *Ezekiel*, London: SCM Press, p. 57.

Zimmerli points out that the word הַנּוֹרָא ('terrible') is not seen in 𝕲 and is part of a later interpretation. Above the ice or crystal firmament was a sapphire throne *kĕmar'ēh ēben sappîr* (כְּמַרְאֵה אֶבֶן-סַפִּיר) (Ez 1:26; 10:1). In Ex 24:10 when YHWH appears to Moses and the seventy elders of Israel, he is standing on a pavement of 'sapphire stone', *ma'ăśêh libnat hassappîr* (כְּמַעֲשֵׂה לִבְנַת הַסַּפִּיר) which is described as being 'like the very heaven for clearness'. A sea of glass like crystal is seen in front of the throne of the one seated on the throne in Re 4:6. These descriptions are similar to the sapphire of the throne and the crystal firmament in Ez 1:22.

If the firmament of Ez 1:22 is an analogy of a firmament that divides the heavens from the earth, this would suggest that although the Living Creatures have wings and go between the earthly and heavenly realms, they are more earthly than heavenly,[491] and more like humans than angels. As suggested above, they may give insights into how heavenly truth is sought out within the earthly realm.

Discussion

If the appearance of *ḥaśmal* is an analogy for the heart, it is also a sign of the love that YHWH has for his people. The *ḥaśmal* is therefore a pivotal symbol in Ezekiel. It connects the Chariot of YHWH image with the crucial phrase from Ez 36:26 'and I will take away from you your heart of stone and give you a heart of flesh', (This is discussed further in 'The interaction of Reason and Spirit'). A connection can also be made between the fire below the loins representing *eros* and Ezekiel equating the breaking of the marriage covenant with that of breaking YHWH's covenant with Israel. The Book of Ezekiel is unique in the Bible in its focus on castigation of Israel's 'promiscuity' (*zānâ*).[492] This

---

491   Launderville, D. (2004), 'Ezekiel's Throne-Chariot Vision: Spiritualizing the Model of Divine Royal Rule,' *Catholic Biblical Quarterly*, **66**, p. 374.

492   A third of Biblical mentions of the word 'harlot' (35/111 King James Version) are in Ezekiel.

theme is seen in both Chapters 16 and 23. In Chapter 16, there is more stress on compassion for Israel because of her hopelessness. God lavishes his love on her which is not only *agape* but also *eros* because Israel came to sexual maturity (Ez 16:7) and YHWH marries her,[493] (Ez 16:8). In Ez 16:19, Israel is said to snub YHWH by running after other gods and sacrificing 'your sons and daughters which you had borne unto me' *bānāyk waĕt-bĕnôtayk ašer yalădt-lî* (בָּנַיִךְ וְאֶת־בְּנוֹתַיִךְ אֲשֶׁר יָלַדְתְּ לִי). In the concluding two verses of Chapter 23, the evils of adultery and of idolatry are melded: 'and I will put an end to lewdness in the land...And you shall bear the penalty of your sinful idolatry' (Ez 23:48-49). This reinforces that unlawful sexual activity, is 'prostitution' and treated similarly to idolatry.*[494] The fire below the loins may therefore be interpreted as a sign of marital love, as well as a symbol of the covenantal love that YHWH holds for his people.

Conclusion of the chapter

The 'likeness of a man' figure appears to represent the majesty or glory of the Lord. The *ḥašmal* may be interpreted as an analogy of the heart and a graphical representation of Ez 26:36. The fire below the loins may be a sign of covenantal love.

[493]   Block, D. I. (1997), *The Book of Ezekiel Chapters 1–24*, Grand Rapids: Eerdmans, p. 482.

[494]   Block, D. I. (1997), *Ibid.*, p. 764; Hosea and Jeremiah also compared idolatry to adultery: Je 5:7; 13:27. Ho 2-5 (2:13).

Chapter 6

Rationality and Measurement in Ezekiel

Overview of Chapter

This chapter looks for evidence of rationality in Ezekiel and the Chariot Vision. Measurement in the Bible and Ezekiel and the potential of the wheel as a metrological tool are reviewed. Rational organisation is discerned in the text of Ezekiel and in the visions including inftances of superior and inferior reason. Mathematics and numerology are reviewed in the Bible and Ezekiel. Babylonian and Ancient Egyptian mathematics and pre-science are reviewed with reference to Ezekiel including the scientific and technological aspects of the wheel. Ezekiel's Vision of the Water Flowing from the Temple is compared with Archimedes' water experiment.

As seen in the Middle Tier of the Chariot (Chapter 3), the Living Creatures appear to conftitute a model for rational decision-making and their movements an analogy for right-reason. If rationality is present in the Chariot Vision it may be found elsewhere in Ezekiel too. Ezekiel uses measurement frequently, especially in the Idealised Temple Vision.[495] This chapter will review measurement in Ezekiel as an indicator of rationality as well as look for

---

[495]   Ezekiel is told 'Let them measure the perfeɕion' (of the temple) וּמָדְדוּ אֶת־תָּכְנִית (Ez 43:10).

rational structures within the text of Ezekiel that might reveal a rational or a 'pre-scientific' bias in Ezekiel's writing. This will include exploration of the wheel in the Chariot for giving 'measurable revelatory knowledge' (see 'Lower Tier': Discussion). Secondly, measurement is one of the most central actions of scientific method and allows a scientific explanation for what is observed.[496] The rational disciplines of mathematics and 'science' will be investigated in the *Sitz im Leben* of Ezekiel and the effect of this background to any rational structure and pre-science content of the book of Ezekiel.

Measurement in the Bible

Measurement (metrology) is the assigning of a standard unit of dimension to an unknown quantity. It is the first and key step in understanding and ordering the natural world.[497] The Bible contains numerous measurements, (particularly in the books of Numbers and of Ezekiel) but biblical metrology is not an exact science. The first measurement units used were naturally-obtained length units such as a man's handbreadth or forearm measurement (a cubit) or the volume of an ass-load for dry or wet measure – (the *homer*). These were broad indications of certain orders of magnitude, but imprecise. The natural or imprecise cubit was used to give a broad idea of the size of an object. For example, a man of great stature in 1Ch 11: 23 was five cubits (7 ft 6 inches) high. Since this figure is rounded to five cubits, it could well be an estimate from a distance. Another example is the height of the gallows used to hang Haman (Es 7:14). It was 50 cubits (75 feet) tall which sounds excessive for the height of a gallows but may have been used to accentuate the evil of Haman.

---

496   Klee, R. (1997), *Introduction to the Philosophy of Science*, Oxford: Oxford University Press, p. 4.

497   Klee, R. (1997), *Ibid.*, p. 4.

An architect or builder needs more accurate measurements. In the Bible, however, it is very unusual to find objective values of different measures.[498] One of the only exceptions is in Ezekiel (Ez 45:11-12) where one term of measurement is defined in terms of another and the *homer* is made the standard measure. The establishment of ratios between measures was a step toward making weights conform to a common standard.[499] A known relation of one measure to another increases accuracy but the actual value of these weights and measures in relation to our modern values is important and comes from a correlation of evidence from a variety of sources.[500] Values from Ezekiel (in Ez 40:5) for length (long cubit and reed), and in Ez 45:11,12,14, for weight (*gerah*), liquid measures (*homer* and *cor*) and dry measures (*homer*) are given in Appendix II in relation to other Biblical measures and their modern equivalents.

In reading measurements in the Bible therefore, it is important to look at the context to see whether the measurement made is (a) meant as a metaphor, a guidance or a moral teaching. (b) an estimate not to be taken accurately or (c) a careful measurement made with the use of a measuring tool. For example, Goliath's height is given as six cubits and a span (1Sa 17:4) which would be 9 feet 9 inches. This is probably a rough estimate rather than an accurate measurement.[501] On the other hand, the cubit measurements of the Idealised Temple Vision in Ez 40-43 were meant to be accurate. It is likely that Ezekiel goes to the trouble of describing each measurement as it was made by the man with a measuring reed in his hand (Ez 40:3), to underscore the accuracy of the measurements being made. Ezekiel also stipulates that the man uses the 'long'

---

[498] Scott, R. B. Y. (1959), 'Weights and Measures in the Bible,' *The Biblical Archaeologist,* **22** (2), pp. 21–40; p. 24.

[499] Scott, R. B. Y. (1959), *Ibid.,* p. 23.

[500] Scott, R. B. Y. (1959), *Ibid.,* p. 22.

[501] Scott, R. B. Y. (1959), *Ibid.,* p. 37.

cubit which is seven palms rather than the normal six in length*[502] (Ez 40:5) (see Appendix 11).

The following are a summary of the main uses of measurement in the Bible.

1. To make an objective, accurate delineation of size of measurable objects:

   'The carpenter stretches a line, he marks it out with a pencil...and marks it with a compass' (Is 44: 13). The measurements of the Ark (Ge 6:15) are in this category.

2. Accurate weights are a means of prevention of dishonesty and cheating:

   'you shall not have in your bag two kinds of measures, a large and a small... A full and just measure you shall have...All who act dishonestly are an abomination to the Lord your God. (De 25: 13-16). Also 'You shall do no wrong in judgment, in measurement of weight, or capacity.' (Le 19:35); Le 19:37: 'You shall have just balances' is echoed by Ez 45:10 'You must use honest scales'.

3. As a metaphor for the measurement of the perceived size of eternal or seemingly unmeasurable things: For example, the Lord's anger (Is 64: 9). 'Do not be angry beyond measure, O Lord,'.

4. To illustrate the Lord's omnipotence: God can measure things that man cannot:

   'Where were you when I laid the foundations of the earth...Who determined its measurements?...Or who set the line upon it?' (Jb 38:5); 'Who has measured the waters in the hollow of His hand, And marked off the heavens by the span, And calculated the dust of the earth by the measure, And weighed the mountains in a balance And the hills in a pair of scales?...' (Is 40:12-16).

---

[502]   The 'long' cubit is 21 inches rather than 18 inches in length.

5. As a metaphor for justice, righteousness, morality, goodness or lack of it:

For example, 'behold I am setting a plumb line in the midst of my people Israel' (Am 7:8); 'common folk are only a breath, great men an illusion, placed in the scales they rise, they weigh less than a breath' (Ps 62:9); 'He stood and measured the earth' (Hb 3:6).

6. To give an eschatological emphasis:

The measurement of the Idealised Temple Vision takes up four chapters in Ezekiel (Ez 40-43). Ezekiel also describes measurements of the future holy district (Ez 45), measurements of the tribal portions of land and of a new Jerusalem (Ez 48). Zechariah also has a vision of a surveyor planning to measure a new Jerusalem, (Ze 2:2) and Revelation (Re 21:16) gives the vast measurements of the New Jerusalem. This use of measurement seems to be a special case of 'spiritual' measurement. The Bible usually proscribes measurement of anything that is spiritual or eternal, as it is, by definition, unmeasurable. The realization of the kingdom of God on earth seems to be an exception to this rule, however, as although it will be a spiritual kingdom, it is for humans who are currently in a finite world. These eschatological measurements seem to have several characteristics: (a) They can be very inflated, like the dimensions of the New Jerusalem (Re 21:16) 2,280 kilometres long; (b) they can be very systematised like Ezekiel's Idealised Temple Vision- symmetrical and square; (c) they are measured by a man with a measuring line (Ez 40:3; Ze 2:1; Re 21:15). In Revelation, the man with a measuring rod has a rod of gold, stressing the importance of measurement.

7. In relation to God- something impossible for man or an attempt to measure the unmeasurable.

'Thus says the Lord, "If the heavens above can be measured ... Then I will cast off all the descendants of Israel for all that they have done,"' (Je 31: 37)

Measurement in Ezekiel

Many measurement values are meticulously documented in Ezekiel, particularly in the final section, the 'Torah' of Ezekiel (Ez 40–48). The following are instances of different measurements that are made in Ezekiel:

1. Frequent measurements of the dates of the writing of the text.

2. Measurement of the number of days the prophet had to lie on each side (Ez 4:5,6) (to equate with the number of years of the iniquity of Israel).

3. The weight of food (Ez 4:10) and measure of drink (Ez 4:11) he had to consume, as a prophecy that the people of Jerusalem would have to measure their food and drink (Ez 4;16).

4. The use of a measuring balance to divide his cut hair into three parts (Ez 5:1).

5. In the Vision of Water Flowing from the Temple (Ez 47:1), measurement is prescribed for the distances where Ezekiel has to cross by cubits, and the depth of the river is measured on his body.

6. The final eight chapters of Ezekiel (the 'Torah' of Ezekiel) are replete with measurements of the temple, and of land apportioned to the tribes of Israel. Of the three sections of the 'Torah': Ez 40:3– 43:27 contains multiple measurements of the dimensions of the temple. The second section, Ez 44:1–46:24 incorporates land areas prescribed for the Levites (Ez 45:1–8), and the third section (Ez 47:1–48: 35) contains the measurement of the river flowing from the temple, and the measurement of Ezekiel's new Jerusalem.

The uses of measurement in Ezekiel can be divided into four main categories:

1. Ezekiel dates his oracles very frequently (fourteen times) and gives year, month and date which is very unusual among the prophets.[503] We can get a much clearer chronology of the historical background to Ezekiel's work as a result.[504]

2. The use of practical day to day measurement: to define just weights and measures (Ez 45:10–12), the proper measured size of an offering (Ez 45:13–15) and also use of measurement in specific situations (Ez 4:10–11 and Ez 5:1).

3. The use of measurement to give a prophetic vision of Israel: Its new temple, the apportioning of land areas prescribed for the Levites (Ez 45:1–8) and to the twelve tribes (Ez 47:13–23 and Ez 48:1–14, and the dimensions of Ezekiel's new Jerusalem (Ez 48:15–35).

4. As part of the Vision of the Water flowing from the Temple. In the vision (Ez 47:3–6), equal measurements are made of the distance down the expanding river, solely to allow the conclusion that the river is expanding exponentially, and not in a rationally predictable fashion.

In summary, Ezekiel uses measurement frequently and the final section of the book contains a protracted measurement section that is seen nowhere else in the Bible. There are 101 measurements made by the man 'with the measuring reed in his hand' (Ez 40:3) between Ez 40:5 and 47:3. The measurement in each case is not only documented, but also the action of measurement with a measuring reed is mentioned every time. The attention to the detail of measurement and its application to varying situations, underlines its importance.

---

[503] Block, D. I. (1997), *The Book of Ezekiel Chapters 1–24*, Grand Rapids: Eerdmans, p. 26.
[504] For example, Ez 1:1 starts 'in the thirtieth year in the fourth month on the fifth day.'

Three questions arise about measurement in Ezekiel: Why is there such meticulous attention to measurement - to defining just measures, to specify measurements of food, drink, time, length. Is it a part of Ezekiel's overall message? Does the frequent use of measurement in Ezekiel have a moral meaning? Was Ezekiel making a statement about the importance of objective observation by measurement rather than relying totally on subjective observations?

In terms of the first question, it would seem that measurement is part of Ezekiel's message. He takes care not only to define just measures but to measure carefully so that it seems the accuracy and repeatability afforded by measurement are attractive to Ezekiel. Having said this, some of the specific values and sizes of the individual parts of the new temple do not seem to be crucial for earthly construction. Also, the measurements of the temple were not made by Ezekiel but by a man dressed in linen under the direction of YHWH, during a Vision, Ezekiel only recorded them meticulously. This new temple, however, is not meant to be built. It is as Block says, a message of hope for the restoration of Israel and YHWH as their God.[505] Of course the overall size of the building is meant to be imposing, but the individual measurements are never used to build it. The more important fact seems to be that the temple is not of arbitrary size, and is not just a vague vision, it is ordained in every detail and in size, by God. Measurement seems to relate more to reflecting YHWH's perfection and that it is a method of bringing his perfection to earth. So the act of measurement becomes for Ezekiel, a means of making YHWH's wishes known to man. Ezekiel seems to use measurement as a 'drawing closer to God'. It is therefore part of his message.

The second question as to whether the measurements have a moral meaning depends partly on what is being measured. It is clear from other parts of the Old Testament, that measurement is used for moral values. The use of the weighing scales as a tool of judgment is one obvious example (Jb 31:6 'Let me

---

505   Block, D. I. (1998), *The Book of Ezekiel Chapters 25–48*, Grand Rapids: Eerdmans, p. 746.

be weighed in a just balance and let God know my integrity'). The weight of food (twenty *shekels*)*[506] and drink (sixth of a *hin*)*[507] prescribed to Ezekiel (Ez 4:10–11), is below a man's daily body requirements and would lead to weight loss and dehydration. The Lord says this is to show what will happen to the inhabitants of Jerusalem who will 'waste away under their punishment' (Ez 4:17). The measurements here contain a moral message and are the numerical equivalent of saying that the people will die for their apostasy. The dividing of his hair into three parts using a balance carries a similar dire warning (Ez 5:12) but in most cases Ezekiel does not directly use measurement as a calibrator of morality.

For the third question, it seems that Ezekiel thought of measurement as being able to raise descriptive language to a higher level of truth than that of empirical observation by itself. This is true for day-to-day measurements, for sacrificial weights and for the measurements of the temple. In this he does begin to think in a pre-scientific way. This would reinforce the importance he places in measurement and why he uses it so frequently.

Metrology and the Wheel

Although the measuring stick predated the wheel, for measurement,[508] there is suggestive evidence that the wheel (the trundle-wheel) was in use for measurement in Ancient Egypt. The trundle-wheel is a rolling disc attached at its centre to an arm so that the wheel can be held by this and rolled along the ground. With a simple trundle-wheel, a mark is made at one point on its circumference and an unknown distance on the ground is measured by running the wheel along it and counting the rotations of the trundle wheel. The distance

---

[506] Equivalent to 450 grams or 0.5 pounds.

[507] Equivalent to 950 milliliters or two fifths of a pint.

[508] Kibble, B. (2014)- The measuring stick came before the wheel, *IEEE Instrumentation & Measurement Magazine*, **17**, pp. 28–29; p. 28.

travelled is the number of rotations multiplied by the wheel circumference or by the diameter of the wheel multiplied by the constant, π.*[509] Trundle-wheels are widely used nowadays but have automatic rotation counters. A trundle wheel of diameter of one cubit has a circumference of π cubits. The length of the base of the Great Pyramid (2,600 BC) is exactly 140 π cubits (or 439.8 cubits).[510] This is strong presumptive evidence that the trundle-wheel was in use for measurement at the time of Ezekiel's prophecy in 592 BC.

Comment and Summary

In the Bible, measurements have several potential applications and meanings. The major dichotomy of use of measurement in the Bible is that it is used to distinguish between things of the earth that are measurable and things of the heavens or spiritual things which are not. The Bible also uses measurement (particularly a plumb-line or scale) for a sense of reckoning or justice. In relation to the abstract world, it can be used metaphorically for the severity of anger for example. The Bible uses the immeasurability of the spiritual to signify God's omnipotence and man's humble position.

Ezekiel knows the importance of measurement and measurements are central to his book. It is the only book of the Bible that gives a section on weights and measures and also gives a cross reference for one measurement in terms of another. The review above also shows that a wheel has potential metrological significance and in the context of the importance that Ezekiel gives to measurement, the wheel at the base of the Chariot Vision could be an analogy for measurement. Ezekiel may have used measurement to raise descriptive language to a higher level of objectivity. Since measurement is at the core of scientific method, and often entails the use of mathematics, Ezekiel's frequent

---

[509] Π (pi) is a mathematical constant which is the ratio of the circumference to the diameter of a circle. The value of pi can be obtained by mathematical calculation. The value of pi can be approximated by the ratio 22/7.

[510]   Hebra, A. (2014), *The Physics of Metrology*, New York: Springer, p. 1.

resorting to measurement may also suggest he has a pre-scientific mindset. Whether the contemporary surroundings of Ezekiel could have fostered this is investigated below.

Evidence of Rationality in Ezekiel

The Book of Ezekiel shows 'a logical development of a series of ideas in accordance with a well thought-out and in part, quite schematic plan.'[511] As noted above, the written word allows a much tighter and argued presentation than an oral one. If we accept that Ezekiel himself wrote the book, it is important to try and determine what place, if any, the writer's own conscious efforts played in the production of the final text. When a vision is complex, there may be considerable psychological input from the visionary before it is committed to paper.[512] Davis sees a highly conscious self-representation in Ezekiel's sign actions:

> a literary effort which is calculated in the best sense: one so finely
> coordinated and attuned to his audience's perception of reality
> that the imitation of an action is more persuasive than the action
> itself.[513]

An important aspect of a written work is that the writer has to 'make' his readers by revealing to them what they have never seen before and moving them to a new order of perception and experience.[514] Written text, more than the spoken word, requires explicit syntactical and lexical features, since a speaker's presence contributes to clarity. However it is the indeterminacies of the text

---

[511] Smend, R. (1880), *Der Prophet Ezechiel*, Liepzig: S. Hirzel, p. xxi.

[512] Andrae, T. (1926), *Mystikens psykologi* (The Psychology of Mysticism), Stockholm: Diakonistyrelsen, p. 117.

[513] Davis, E. F. (1989), *Swallowing the Scroll: Textuality and the Dynamics of Discourse in Ezekiel's Prophecy*, Bible and Literature Series 21, Sheffield: Almond, p. 71.

[514] Booth, W. (1983), *The Rhetoric of Fiction*, Chicago: University of Chicago Press, p. 396.

that engage the reader and 'indeterminacy is the fundamental precondition for reader participation.'[515] The writer has to supply the logical gaps in these indeterminacies and it is in supplying these gaps that the reader gives a personal input.[516] A logical rational approach is essential in conveying ideas to a readership, particularly if the ideas are complex, novel and at times bordering on the bizarre.[517] This section will look for evidence of rationality within Ezekiel, both with the Chariot Vision as well as in the text of the book of Ezekiel. It will firstly look for rational organisation within Ezekiel and other evidence of rationality. It will explore different levels of rationality: superior reason and inferior reason. Thirdly it will examine whether the wheel could be a link between rationality and 'science'.

Rational Organisation in the Book of Ezekiel

The book shows three major underlying strands of organisation:

1. The first is the Chariot Vision whose repeated appearance tells a rational story: the 'Glory of God' appears at the start of the book (Ez 1:26), it goes out from the Temple (Ez 11:22) after the Vision of Abominations in the Temple (Ez 8:1-18) and then it returns to the temple when the future reconstructed temple is measured (Ez 43:4).

2. The second organisational structure is the ability to divide the final Idealised Temple Vision (chapter 40-48) into three sections of comparability

---

515   Iser, W. (1971), 'Indeterminacy and the Reader's Response in Prose Fiction,' in *Aspects of Narrative,* Miller J. H. (ed.), New York: Columbia University Press, p. 14.

516   Davis, E. F. (1989), *Ibid.,* p. 74.

517   For example, Ezekiel puts forward new laws about personal responsibility (Ez 3:16–21; 18:1–32). These are an emendation of a prior existing Mosaic law: that responsibility is borne by offspring for even several generations (Ex 34:7). In putting forward this law Ezekiel rationally challenges Mosaic law: a child has a separate will and conscience from their parent and therefore they should not be bound by parental sins, (This had relevance for Ezekiel's time as in Ez 11:15 the Exiles were said to have been rejected by God because of their ancestors' sins).

with the Mosaic Torah: (i) YHWH coming to reside in the temple; (ii) Israel's response to YHWH's presence; and (iii) apportioning of land to the twelve tribes.[518] (see section 'The Torah of Ezekiel')

3. The third is the division into three themes as noted above: Oracles of judgment; Oracles against foreign nations; Oracles of salvation.

Further evidence of rational arrangement can be seen in the oracles. The metaphorical oracles in Ezekiel appear to be organised, and the metaphors appear to be used in a systematic fashion.[519] Individual oracles are often divided, with a correspondence between each half. These structural features suggest deliberate design.[520] Two subunits are also seen in Ez 37 which is skilfully worked in its literary context:[521] the basic themes of exodus, return and reunification are seen both in the Vision of the Dry Bones subunit (Ez 37:1-14) and in the Two Sticks subunit (Ez 37:15-28).[522,523] These two complementary units show a rhetorical genius in Ezekiel that is both logical and rational.[524]

Other Evidence of Rationality

Ezekiel is very careful to qualify many of his statements in describing the Chariot Vision which has many 'indeterminate' features. He frequently uses 'appearance of' and 'likeness of' when describing the Chariot Vision, which

---

518 Block, D. I. (1998), *The Book of Ezekiel Chapters 25–48*, Grand Rapids: Eerdmans, p. 590.

519 Schöpflin, K. (2005), 'The Composition of Metaphorical Oracles within the Book of Ezekiel,' *VT* **55** (1), 101–120; p. 101.

520 Block, D. I. (2013), *By the River Chebar: History, Literary, and Theological Studies in the Book of Ezekiel*, Eugene: Cascade, p. 6.

521 Odell, S. (2005), *Ezekiel*, Smyth & Helwys Bible Commentary, Macon: Smyth & Helwys, p. 451.

522 Barth, M. (1984), 'Traditions in Ephesians,' *NTS,* **30** (1), p. 3.

523 Suh, R. H. (2007), 'The use of Ezekiel 37 in Ephesians 2,' *JETS,* **50** (4), pp. 715–733; p. 721.

524 Block, D. I. (1998), *Ibid.,* p. 267.

was a complex moving vision (Ez 1:27-28). This shows that he does not want to mislead by making a conclusion with insufficient evidence. This contrasts with the precise detailed measurements of the Idealised Temple Vision, where he does not use similar qualifying phrases.

As noted above, since measurement is the assigning of a standard unit of dimension, it is rational. In Ez 43:11 Ezekiel is told to make known 'the fashion/delineation/design of the temple and its measurement/proportion': *ṣûrāt habbayit wĕtĕkûnātô* (צוּרַת הַבַּיִת וּתְכוּנָתוֹ). When we measure what has been observed, the numbers or mathematical entities produced, allow confirmation of the observation. In Ezekiel the Idealised Temple and Water Flowing from the Temple Visions incorporate multiple interdependent measurements which give a rational basis to the visions and are part of the rational organisation in the book.

In summary, Ezekiel used rational structures to organise his book. He also uses indeterminate information in a careful rational way and even the visions contain rational structuring. There is therefore evidence of rational organisation in both in Ezekiel's writing and in his visions.

Superior versus Inferior Reason

In Chapter 3 it was argued that the horizontal movements of the Living Creatures are an analogy for right reason, since the Chariot moved straight without turning. It was shown in Chapter 4 that the wheels' straight movements are also due to the suffusion of the Living Creatures with the spirit since the movements of the Living Creatures and those of the wheels are bound together through the spirit. Since the spirit of the Creatures directs the wheels, the 'rationality' of the spirit-endowed movement of the Creatures therefore is likely imparted to the wheels. The Creatures' part angelic and part human anatomy link heavenly and earthly realms. The wheels however are hierarchically below the Creatures, in that wheels are an invention and a construction of man,[525] but are divinised by the fact that they are covered with eyes.[526] If they are also imbued with the rational spirit of the Living Creatures, like the Creatures, they must be serving a function in the chariot in relation to rationality. Although the wheels' movements are tied to the Creatures through the spirit, the fact that wheels are fundamentally of earthly origin, suggests that they may have a different revelatory focus to the creatures. It was argued in Chapter 3 that the Living Creatures are a model for rational decision making. Although the wheels may take part in the rationality of the Creatures, it is an 'earthly' rationality compared to that of the Creatures, as the wheels contact the ground. The Creatures' rationality is transmitted by the four character-forces portrayed by the Creatures. The wheels rationality is transmitted by its rigid constructed structure.

Augustine (supported by Aquinas) divided reason into superior and inferior: Superior reason is fixed on eternal things and draws wisdom from them, inferior reason is turned towards exterior, material things and through knowledge ('science'), that it acquires, guides man through the practical affairs of life.[527] The Living Creatures appear to enfold both superior and inferior reason but the earthly construction of the wheels suggests that they are involved solely with inferior reason.[528] In order to better understand the interaction between the wheels and the Living Creatures, this subsection will look more specifically

---

[525] The wheels are the only part of the vision that incorporates artificial construction, Cf. Eichrodt, W. (1970), *Ezekiel*, London: SCM Press, p. 57.

[526] Multiple eyes are well known from Assyrian statues of gods and reveal the all-seeing divine presence. Cf. Boadt, L. (1990), *Ezekiel*, New Jerome Biblical Commentary, London: Burns & Oates, p. 310.

[527] Augustine, *De Trinitate*, XII: 3, 'the lower part of the soul is turned towards exterior, material things, and, through the science it acquires, guides man in the practical affairs of life'; Aquinas, *Summa Theologiae*, 1:79:9; Moya Cañas, P. (2013), 'Ratio superior, ratio inferior y el agustinismo de Tomás de Aquino,' *Anuario de Estudios Medievales*, **43** (2), pp. 777–798, p. 784.

[528] Gilson, E. (1956), *The Christian Philosophy of St Thomas Aquinas*, Indiana: University of Notre Dame Press, p. 346, says 'to the extent that human reason which always remains one and the same, is applied in the acquiring of science, it is called "inferior reason".'

at instances of apparent superior or inferior reason in Ezekiel and whether any interactions between them that may be discerned.

Although there is a logical development of ideas in Ezekiel,[529] for most of the rational structures in Ezekiel, it is difficult to determine whether they are based on superior reason, inferior reason or a combination of both.*[530] For example, as mentioned above, there is rational structure in the themes of the oracles, and there is also rational organisation within the oracles. Although these constructs appear to be 'practical' and therefore inferior reason, they do relate to spiritual issues with implied pursuit of wisdom, and so cannot be clearly classified as superior or inferior reason.

The weight standardisations introduced by Ezekiel are clearly inferior reason applications.[531] The individual measurements that are performed in the Visions of the Water Flowing from the Temple (Ez 47:3) and in the Idealised Temple Vision (Ez 40:5) are also measurements which are rational as they are part of delineation of a larger plan. Each measurement is an externally orientated application of reason and would represent inferior reason.

Instances of superior reason are seen principally in the visions: The repeated appearances of the Chariot Vision through the Book of Ezekiel and their organisation, tell the story of YHWH's holiness, his response to Israel's idolatry and how He will return with compassion despite their sins.[532] The story behind the Chariot Vision's successive appearances is superior reason. The Idealised Temple Vision is primarily a rational theocentric exposition of Israel's future,[533] and it also would fit with superior reason.

---

[529] Smend, R. (1880), *Der Prophet Ezechiel*, Liepzig: S. Hirzel, p. xxi.

[530] Inferior reason is applied in acquiring science (knowledge) and Superior reason in attempting to acquire wisdom. St Augustine, *De Civitate Dei*, 40 27:2.

[531] Block, D. I. (1998), *The Book of Ezekiel Chapters 25–48*, Grand Rapids: Eerdmans, p. 657.

[532] Block, D. I. (1998), *The Book of Ezekiel Chapters 25–48*, Grand Rapids: Eerdmans, p. 590.

[533] Block, D. I. (1998), *Ibid.*, p. 746.

Interactions between superior and inferior reason are seen on two occasions: The Idealised Temple Vision deals directly with the divine, so it is formulated with the use of superior reason, but the individual measurements are at an inferior level. Similarly, in the Vision of Water Flowing from the Temple, the individual measurements made can be said to be inferior reason, as they are a simple interaction with the environment. However, the final rhetorical question that YHWH asks Ezekiel: 'Son of man, have you seen this?' (Ez 47:6), is asking Ezekiel to use superior reason and to make sacred inferences from the conclusions he reaches from exercising inferior reason.

In these last two instances we see that inferior reason establishes the basis for a move to superior reason. These two visions are rational structures that have measurement of concrete objects (within visions however) as their inferior reason foundation. This repeated measurement establishes a pattern or structure which when submitted to superior reason gives divine insights. It supports an interaction between superior and inferior reason that echoes the close linkage of the Living Creatures and wheels working through the spirit.[534] Inferior reason can therefore be instrumental in reaching a superior reason conclusion. This exemplifies the maxim that reason has the capacity to deliver a higher level of objective truth.[535]

Mathematics and Numerology

It was suggested above that the frequent use of measurement in Ezekiel could suggest he had a 'pre-scientific' mindset. If pre-science is present in Ezekiel, it is important to see if this reflected what was going on in his *Sitz im Leben*. Mathematics is a building block of science, and like science, is a rational dis-

---

534   Launderville, D. F. (2007), *Spirit and Reason,* Waco: Baylor University Press, p. 94.
535   Launderville, D. F. (2007), *Ibid.,* p. 348.

cipline and predated science historically.[536] The Bible gives a specific meaning to certain numbers, (for example a period of trial or preparation lasts forty days) and older societies have incorporated numerological meanings in certain numbers. Contemporaneous development of mathematics and numerology will be reviewed here for any connections with Ezekiel. Evidence of mathematics and numerology will also be looked for in the text of Ezekiel.

Numerology

Although numbers are abstract entities there are indications that numbers are often embedded in nature. For example, the shape of a snail's shell and the numbers of petals in different flower species are based on the Fibonacci sequence (a sequence in which each number is the sum of the two preceding numbers starting from 0 and 1).[537] Fibonacci sequence numbers are seen for example, in biology where five is seen frequently: from the number of sections in an apple to the number of leaves on a tomato stalk and number of stamens in the passion fruit flower. Mammals have five digits on each limb. Eight is also common: spiders have eight legs and octopuses have eight legs. The fact that numbers in the Fibonacci sequence are frequently found in nature suggests that these numbers have a deeper meaning than just being abstract quantitative markers.[538] The 'golden ratio' (1.618)*[539] denoted by the Greek

---

[536]   Clagett, M. (1999), *Ancient Egyptian Science: A Source Book*, Volume 3, Philadelphia: American Philosophical Society, p. 1; Mathematics was used in Egypt as far back as 3000 BC, Early or pre-science as defined by evidence of use of inductive reasoning in predictive algorithms in astronomy was present in Babylonia in about the 8th century BC: Brown, D. (2000), *Mesopotamian Planetary Astronomy–Astrology*, Groningen: Styx Publications.

[537]   The Fibonacci sequence starts 0, 1, 1, 2, 3, 5, 8, 13, 21, 34, 55...Daisies usually have 34 or 55 petals; marigolds 13, buttercups 5.

[538]   Fanthorpe, N. (2013), *Mysteries and Secrets of Numerology*, Toronto: Dundurn, p. 77.

[539]   A way to calculate the Golden Ratio is using the square root of 5. The square root of 5 is approximately 2.236068, so the Golden Ratio is approximately $0.5 + 2.236068/2 = 1.618034$.

letter phi (φ) is also seen in nature. Two quantities are in the golden ratio if their ratio is the same as the ratio of their sum to the larger of the two quantities. Buildings whose outer dimensions contain this ratio, are aesthetically pleasing. The golden ratio is seen in the proportions of a snail's shell which also contains the Fibonacci sequence.

Egyptian Numerology

The three sides of a right-angled triangle have a fixed ratio of 3: 4: 5. The Egyptians used this ratio in building.[540] This ratio, which they used to construct all other geometric figures and three dimensional solids, was fundamental for Egyptians. The number 3 was also used to represent a triangle, 4 to represent a square and 5, a pentagon.

The Egyptians liked to arrange their gods in threes for example *Atum, Shu* and *Tefnut*; *Horus, Isis* and *Osiris. Ra*, the sun-god was shown in three ways: at dawn, midday and sunset. *Thoth* was known as thrice-blessed and was 'thrice-great'.

The Pyramids and the Mathematics of a Circle

There is an interesting relationship between the construction of the pyramids, particularly the Great Pyramid of Khufu at Giza, and the circle or sphere. It was found that the height of the Great Pyramid is the dimension of the radius of a circle whose circumference is the dimension of the square perimeter of the base of the pyramid.[541, *542] It appears that the proportion of 7:22, which is a close approximation of the value of the mathematical constant *pi* (π), has

---

540 Fanthorpe, N. (2013), *Ibid.*, p. 108.

541 Petrie, F. (1883), *The Pyramids and Temples at Gizeh*, London: Field & Tuer, p. 220.

*542 It is important to clarify, however, that the base of the pyramid is a square and that if it was a circle the 3 dimensional shape would be a cone, not a pyramid.

been purposely built into the dimensions of the pyramid.[543] If you divide the square perimeter of the great pyramid by its height you get $\pi/2$. In addition, the slope of the faces of the great pyramid are very close to $\pi/4$ (0.785). Petrie found that the walls of the king's chamber within the pyramid are determined by the same π proportion as the exterior of the pyramid. Moreover, the length of a side wall of the King's chamber is equal to the diameter of a hypothetical circle the circumference of which is the distance of the circuit round the four straight dimensions of a side wall of the chamber. The width of the King's chamber is also equal to the radius of this hypothetical circle. Taylor suggested that the pyramid was built in this way in order that it may be a representation of the spherical earth.[544] This does not however explain why the π proportions are also used in the King's chamber walls and floor. Greenberg is of the contrary opinion that the relationship between π and the Great Pyramid is a coincidence. This is suggested by two facts – one purely mathematical and the other historical, but both involving the number seven: (1) the rational number obtained by dividing 22 by 7 happens to be an excellent approximation to the constant π; (2) the fact that the Egyptian measurement system is based on the number seven, as one cubit is divided into seven equal units (palms).[545]

In addition to the proportions of π, many of the proportions of the great pyramid are close to the 'golden ratio' of 1.618. There is however no written evidence that the Egyptians knew about this ratio or calculated it and the presence of the ratio in the pyramid could be due its proportions.

---

543 Taylor, J. (1859), *The Great Pyramid: Why Was It Built? And Who Built It?*, Cambridge: Cambridge University Press, p. 22; for an explanation of the how pi relates to a circle see the Section 'The Wheel: Analogy for Science and Technology'.

544 Taylor, J. (1859), *Ibid.*, p. 19.

545 Greenberg, R. (2000), 'Pi and the Great Pyramid,' at *https://sites.math.washington.edu/greenber/PiPyr.html* accessed 11/5/2019.

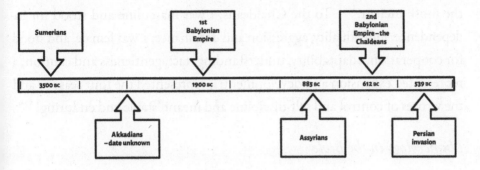

Figure 7: Timeline of Mesopotamian Empires.

Babylonian Numerology

Babylonians used a sexagesimal counting system, which uses a base of 60 instead of a base 10 as we use nowadays. The Babylonian system was unique and had many advantages, particularly for computing. Sixty is easily divisible by 2, 4, 6, 10, 12, 15, 20 and 30 and so the system facilitates mental calculations and circular measurement. We still use this system for time and for degrees of a circle as well as for astronomical location. Seven was an important number as well for the Babylonians: it represented analysis, knowledge, understanding, awareness and study.[546] Gilgamesh, the fifth king of Uruk, Sumeria, (Figure 7) reigned between 2,800 BC and 2,500 BC. He was 'deified' posthumously in the Epic of Gilgamesh[547] which was a mythical epic poem written about him in the late second century BC. The number seven has mythical significance in the story: there is a gate with seven bolts. Seven mountains are climbed. seven cedar trees are cut down. Gilgamesh tricks a giant with seven gifts and there was a flood which lasted seven days. Chaldean numerology is probably

---

546   Fanthorpe, N. (2013), *Mysteries and Secrets of Numerology,* Toronto: Dundurn, p. 129.
547   Pritchard, J. B. (2011), *The Ancient Near East: An Anthology of Texts and Pictures,* Princeton: Princeton University Press, p. 39.

the most ancient.*[548] To the Chaldeans, 1 was masculine and stood for independence, individuality, aggression and dominance; 2 was female, and stood for cooperation, adaptability, understanding, tact, gentleness and caution; 3 represented expansion and development; 4 represented the four seasons and the virtues of control and self-discipline and meant 'stable and enduring'.[549]

Numerology in the Bible

In the Bible we find several numbers that have important spiritual significance. The number 'one' is holy in the Bible and is central to the divine nature and indicates unity. The number two is a symbol of union. This is seen in the two natures of Christ and in marriage. It can also represent division. Number three is a special number for Biblical numerologists[550] as it represents the Holy Trinity, the division of time into past present and future as well as the three dimensions of space.

Four is a number of stability, foundation, order and the completion of justice: the four gospels, the four living creatures of the Apocalypse (Re 5:6), the four horsemen of the Apocalypse (Re 6:2), the four angels who hold the winds (Re 7:1).

Five is said to signify activity, freedom, adventure or change.[551] In the Old Testament David selected five stones with which to defeat his enemy (Sa 17:40). In the New Testament, the wounds of Christ are the most important manifestation of the number five. Six has the meaning of responsibility and importance or balance.[552] In the story of Genesis, humans were created on the sixth day (Ge 1:31) and there were six steps leading up to the throne of King Solomon,

---

[548] Chaldea was a country in the Ancient Near East which was assimilated into Babylonia in the mid sixth century BC.

[549] Fanthorpe, N. (2013), *Mysteries and Secrets of Numerology*, Toronto: Dundurn, p. 50.

[550] Fanthorpe, N. (2013), *Ibid.*, p. 195.

[551] Fanthorpe, N. (2013), *Ibid.*, p. 65.

[552] Fanthorpe, N. (2013), *Ibid.*, p. p. 83.

(1 Ki 10:19). Seven is connected with holiness. God rested on the seventh day and sanctified it (Ge 2:3). It took seven years for Solomon to build the Temple (1Ki 6:38). Blood is sprinkled seven times on the mercy seat on the day of Atonement (Le 16:14) and also sprinkled seven times on the altar to make it holy from uncleanness (Le 16:19). There were seven lamps in front of the lampstand in the tabernacle of meeting (Nu 8:14) and in front of the one like the son of man in the Book of Revelation (Re 1:12).

Forty is also a frequently used number for the passage of time, in terms of trial and hardship. It is the number of years the Hebrews wandered in the desert, (De 8:2) the number of days it rained in the flood, (Ge 7:12), the number of days Moses was on the mountain, (Ex 24:18) and the number of days and nights it took Elijah to walk to the mountain of God, Horeb (1Ki 19:8). In the New Testament, forty was the number of days Jesus fasted (Mt 4: 2) and the number of days Jesus was on the earth after the resurrection. (Ac 1: 3)

Babylonia – Origins of Mathematics

A review of Babylonian mathematics is relevant for this thesis because pre-scientific thought arose within mathematical astronomy in Babylonia and there was an important link between astronomy and religion in Babylonia. Links between the Living Creatures, rationality and the cosmos found above, suggest that it is important to look for further potential links between the Chariot and Babylonian pre-science.

The evidence we have for Babylonian mathematics comes from thousands of clay tablets that were discovered since the 19th century. Starting in about 3,500 BC Sumerians developed a method of writing using wedge shaped (cuneiform) markings in wet clay. Clay tablets are very durable and about five hundred tablets of mathematical interest have been found. Deciphering of the tablets did not start until the late 19th century but by 1920 the methodology of Baby-

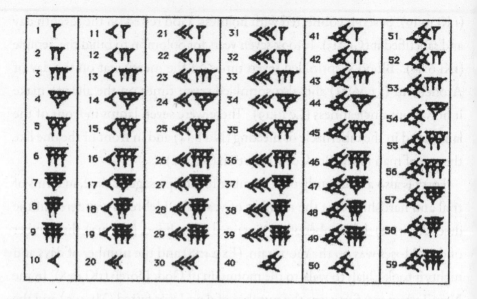

Figure 8: Babylonian cuneiform system of numbers up to 59. Source: *https://history.st-and.ac.uk/HistTopics/Babylonian_numerals.html*.

lonian mathematics was well understood.[553] Mathematics was first developed in Sumeria/Babylonia about 3,000 BC.[554] In 2,000 BC the Babylonians started to use a simple number system with two cuneiform symbols: a pin shape that represented one and a wing shape representing 10. This enables counting up to 60 (Figure 8).*[555]

This was a sexagesimal (base 60), place-value numeric system. The origin of the sexagesimal base is unclear although the Sumerians who practised astro-

---

553 Høyrup, J. (2002), *Lengths, widths, surfaces: A portrait of old Babylonian algebra and its kin*, Sources and Studies in the History of Mathematics and Physical Sciences, New York: Springer.

554 van der Waerden, B. L. (1975), *Science awakening*, Leiden: Noordhoff International, p. 62.

555 The incorporation of a decimal component to Babylonian number notation and counting suggests the number of digits on the hands. 60 may also have been reached by multiplying the lunar months (12) by the number of digits on one hand.

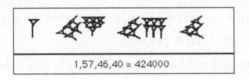

1,57,46,40 = 424000

Figure 9: Raising to powers of 60. Source: O'Connor, J. J., Robertson, E.F. (2000), *Ibid*.

nomy by recording the motion of celestial objects, could have used the number of lunar months in a year (twelve) as a base. This could have been multiplied by the number of visible planets at the time (Mars, Venus, Mercury, Saturn and Jupiter).[556] A more practical and convincing explanation however is that if one uses a right finger to point successively to each of the three natural flexion creases on each of the four fingers (excluding thumb) on the palmar aspect of the left hand one can count to twelve. By repeatedly counting these twelve creases using each of the five digits of the right hand in turn, one can count up to 60. This uses the right-hand digits as multipliers of the twelve left finger creases ($5 \times 12 = 60$).*[557]

The Babylonian sexagesimal system was used primarily as a counting system and allowed systematic accounting using clay counters for traded and stored goods. In 2,500 BC the system was improved by using a place notation, with figures to the left being raised by increasing powers of 10, as in our system of number writing. Cuneiform signs for numbers higher than 60 are grouped in two or more groups with a small space between each group. Each succeeding group to the left raises the digits in that group to progressive powers of 60 (60^1, 60^2, 60^3...). For example (Figure 9) our decimal number 424,000 is expressed as 1,57,46,40 in a sexagesimal system: this is (going from right to left in the

---

556   O'Connor, J. J., Robertson, E. F. (2000), 'Babylonian Numerals,' at *https://mathshistory. st-andrews.ac.uk/HistTopics/Babylonian_numerals/* accessed 22/4/2019.

557   This is an example of using an external scaffold to assist symbolic reasoning. See Section on 'Psychological background to Symbolic Reasoning' in Chapter 7.

figure): [40] + [46 × 60 (=2,760)], + [57 × 60 × 60, (= 205,200)] + [1 × 60 × 60 × 60 (= 216,000)], Total = 424,000.

The Babylonians did not have a symbol for zero and so left a space for zero, to signify that the next figure to the left should be raised to the subsequent higher power of 10. The system was open to ambiguity and so between 700 BC and 300 BC a placeholder symbol of two little triangles was added.[558] Links have been found between the Babylonian sexagesimal system and the Egyptian centesimal system. Friberg found evidence of use of both systems on a single clay tablet suggesting that both systems may have been taught simultaneously in some places.[559] In fact the Babylonian system is a combination of decimal and sexagesimal systems because they counted in tens and in sixes, the pin sign being the number 10. Babylonians made calculations using tables of the values of squares, cubes, reciprocals, square-roots and cube-roots. In addition, they had tables for squares and cubes for integer values from 1 to 20 as well as for the integers 30, 40, and 50.[560]

In the second millennium BC, using this system, Babylonians were able to compute an approximation of the square root of 2.*[561,562] They also had an early form of algebra*[563] and knowledge of what is now called Pythagoras's theorem

---

558   Zara, T. (2008), 'Babylonian Mathematics,' Senior Honors Thesis, Liberty University, Lynchburg, Virginia, p. 12. at: *https://pdfs.semanticscholar.org/7b8b/d86560c8fd94c8ff 35docfcfc77b8ab6bf95.pdf*

559   Friberg, J. (2005), *Unexpected Links between Egyptian and Babylonian Mathematics*, London: World Scientific, p. 3.

560   Teresi, D. (2002), *Lost discoveries: The Ancient Roots of Modern Science – from the Babylonians to the Maya*, New York: Simon & Schuster.

561   Square roots are important building blocks in statistics and quadratic equations. The square root of 2 ($\approx$1.41) is an irrational number that cannot be written as a simple fraction.

562   Robson, E. (2008), *Mathematics in Ancient Iraq: A Social History*, Princeton: Princeton University Press, p. 1.

563   Algebra is defined as a form of mathematics that uses variables, in the forms of letters and symbols, to act as numbers or quantities in equations and formulas.

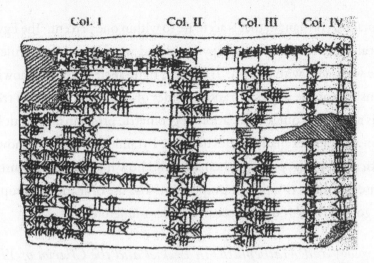

Figure 10: Drawing of Tablet known as Plimpton 322. Source: Van der Waerden, B. L. (1975) *Science awakening*, Leiden: Noordhoff International, p. 78.

of the relationship of the squares of the length of the sides of a right-angled triangle.[564] This is shown in the tablet known as Plimpton 322 (Figure 10).

Egyptian Mathematics

The earliest Egyptian mathematical text discovered is the Moscow Papyrus, which dates from about 2,000-1,800 BC. The Egyptians introduced the earliest well-developed base-10 number system with which they could do all basic mathematical calculations. They were able to use natural fractions (half, thirds, quarters) and unit fractions (which have a numerator of 1). They were able to solve second-order algebraic (quadratic) equations. They discovered that the area of a circle of diameter 9 units was approximately the same as that of a square of sides of 8 units. They used this to calculate the area of a circle by multiplying the diameter by 8/9 and squaring it. This gives a value for the

---

[564] Robson, E. (2008), *Mathematics in Ancient Iraq: A Social History*, Princeton: Princeton University Press, p. 110. Pythagoras's theorem is useful in navigation and can be used to determine the shortest distance between two places.

mathematical constant π, that is accurate to within one percent. The Egyptians could calculate the volume of a pyramid and were aware of the rule that a triangle with sides with ratios 3: 4: 5 gives a right-angle triangle. Knowing the relationships of the angles or ratios of sides of special right-angled triangles like this allows the quick calculation of geometric problems, without having to use more complex mathematics to do so. Pyramid construction showed the sophistication of Egyptian mathematics and how it allowed the planning of immense buildings to a pre-set proportion. The use of π and the proportion of the 'golden ratio' (1.62) is discussed under numerology.

Numerology and Mathematics in Ezekiel and the Chariot of YHWH Vision

In the Book of Ezekiel, the number four is seen in the square shape of the temple in the Idealised Temple Vision (Ez 42:15), the four measurements made of the depth of the river in the Vision of Water flowing from the Temple. (Ez 47:3), and the four winds from which the spirit comes to breathe upon the dry bones in the Vision of the Dry Bones. (Ez 37:9). The Book of Ezekiel has 48 chapters. Zimmerli says that Ezekiel brings out the importance of the number four.[565]

In Ezekiel five appears three times in the first paragraph of the book in the preamble (Ez 1:1). Ezekiel here points out the coincidence that his vision occurred on the fifth day of the month and that it was the fifth year of the exile of King Jehoiachin. Ezekiel's second vision, that of the abominations in the Temple, also occurred on the fifth day of the month (Ez 8:1). The report of the fall of Jerusalem also came to Ezekiel on the fifth day (Ez 33:12). The Idealised Temple Vision started in the 25th year of the Exile on the tenth day of the month. The overall dimensions of the Temple Complex were 500 × 500

---

565   Zimmerli, W. (1979), *Ezekiel 1*, Philadelphia: Fortress Press, p. 120.

cubits. Many of the measurements of the temple are multiples of 5.*[566] The portions of land allotted to the different tribes of Israel were also all multiples of five (Ez 48:15).[567] The number five links Ezekiel to Christ because of Christ's five wounds.[568]

There were six executioners (Ez 9:2) in the Vision of the Executioners.

The word of the Lord came to Ezekiel after seven days (Ez 3:16) and also on the seventh day of the month (Ez 30:20). In the Torah of Ezekiel, there is a connection with Leviticus: the altar had to be consecrated by making atonement for seven days (Ez 43:26). The number seven is repeatedly used for feasts of atonement: it is held on the seventh day of the month for seven days with offering of seven bullocks and seven rams (Ez 45:23). As noted above, there are seven mentions that the Chariot moved straight and did not turn, (See 'Middle Tier'), suggesting that moving straight may have a holy significance.

Eleven in biblical numerology, relates to the eleventh hour, which is a metaphor for when time has nearly run out or the end of times. (Mt 20:1) There are 236 months or 7211 days between the dates of the first and the last appearance of the Chariot Vision. Both these numbers add up to 11. There are 11 levels of sacrality in the vision.[569] Also 101 measurements are made on the temple, this number again suggesting the number 11. This suggests that the Idealised Temple Vision relates to the future, towards the 'end of times'.

The prophet lay on his right side for 40 days to bear the punishment of the house of Judah (Ez 4: 6) and in the prophecy against Egypt it is the number of years for which the Egyptians will be scattered (Ez 29: 11).

The numbers that are obvious in the Chariot Vision are:

---

[566] For example, the depth of the inner temple court and distance between inner and outer gates was 100 cubits. The depth of the inner east gate was 50 cubits. Many walls were 5 cubits thick. See Block, D. I. (1998), *The Book of Ezekiel Chapters 25–48*, Grand Rapids: Eerdmans, Figure 4, p. 541 and Figure 6, p. 565.

[567] Block, D. I. (1998), *The Book of Ezekiel Chapters 25–48*, Grand Rapids: Eerdmans, p.733.

[568] Cf. Christman, A. R. (2005,) *'What did Ezekiel See?,'* Leiden: Brill, p. 25.

[569] Block, D. I. (1998), *Ibid.*, p. 571.

One: the figure like a man who sits on the throne (Ez 1:26); Two: the pairs of wings on each living creature. (Ez 1:6) The doubling of each of the wheels (Ez 1:16).

Four is the dominant number in the Chariot Vision. There were four wheels to the chariot, (Ez 1:15) four Living Creatures, four faces to each living creature, (Ez 1:6) and four wings on each Living Creature (Ez 1:11). In the Babylonian tradition, four represented stability, enduring, and the four seasons. Greenberg says that the dominance of the number four (wheels, Living Creatures, wings) is connected with the division of the world into four parts, or the circle of the horizon into four directions. It symbolises the divine capacity to control the whole world.[570] The wheels and the square shape of the base of the vision give stability to the whole structure. In Ezekiel, the number four has cosmic connotations because of its links with the four winds and the four corners of the heavens,[571] and implies the omnipotence of God. Four also means 'rationality' which links to mathematics.

Conclusion: Numerology and Mathematics in Ezekiel

While there are no clear-cut uses of mathematics in Ezekiel there are several suggestions that Ezekiel was familiar with numbers and probably was familiar with Babylonian mathematics. The central use of the number four in the Chariot Vision underscores the importance of rationality. The frequent use of measurement in Ezekiel implies the importance of mathematics, in applying weights and measures (Ez 45:10). Particularly his establishing of the *homer* as the standard measure and his fixing of the weight of the *shekel* at 20 *gerahs*, suggests that he had to have performed mathematical calculations to compare the existing measures. Ezekiel also changed the size of the *minah* to 60 shekels from the prior Old Testament value of 50 shekels. This was probably inspired by the

---

570   Greenberg, M. (1983), *Ezekiel 1–20*, New York: Doubleday, p. 57.
571   Irenaeus, *Adversus haereses*, III.11.8.

Babylonian sexagesimal system[572] as it is certain that Ezekiel would have to have used this system. The cuneiform script and sexagesimal numbering system were powerful, efficient and easily learned systems of formal mathematical and logical notation, which facilitated the development of Babylonian mathematical expertise.

Pre-Science in Babylonia and Egypt

Sumerians practised astronomy beginning in the second millennium BC and knew of most of the planets. In the Seleucid Era (312 BC) the Babylonians made tabulated predictions of positions of the sun, moon and five visible planets.[573] Up to about the fifth century BC, astronomy was the concern solely of the king and his court and was linked with religion and astrology. In Old World Astronomy heavenly objects were often identified with the gods. The Babylonians however, applied mathematics to astronomy. There is evidence that the earliest attempt at accurate prediction of planetary movement occurred in 8th to 7th century BC Babylon. This has been called the 'First Scientific Revolution'.[574] The Babylonians invented the zodiac by dividing the path of the sun and the planets into 12 equal segments of 30° and naming each sector on the basis of a nearby constellation of stars. This became a coordinate system that facilitated the development of mathematical astronomy.*[575] Babylonian mathematical astronomy was written in cuneiform script on clay tablets from the fifth century BC. Most of the evidence for mathematical astronomy comes from the later first millennium BC, and shows that the Babylonians made

---

[572] Block, D. I. (1998), *The Book of Ezekiel Chapters 25–48*, Grand Rapids: Eerdmans, p. 657.

[573] Robson, E. (2008), *Mathematics in Ancient Iraq: A Social History*, Princeton: Princeton University Press, p. 218.

[574] Brown, D. (2000), *Mesopotamian Planetary Astronomy-Astrology*, Groningen: Styx Publications, p. 6.

[575] Mathematical astronomy is the use of mathematics to count objects in the sky, to calculate distances, or to predict movement of planets.

astronomical observations, and models of calculation that deeply impacted the development of Old World astronomy.[576] Babylonian scholars developed period relations*[577] to describe and predict key celestial events. Babylonian astronomers were successful in predicting lunar eclipses 55% of the time, an additional 35% being near misses. They also predicted the total solar eclipse of 136 BC.[578]

The scientific aspect of astronomy was looking for patterns in apparently random planet movement. Movements of Jupiter were often subjected to mathematical analysis since Jupiter was sacred to Marduk, the supreme god of the Babylonians. Jupiter's relative movement as viewed from earth varies from anterograde to retrograde, stopping in between in a sinusoidal fashion. This pattern occurs because Jupiter is very distant from earth and orbits the Sun more slowly.*[579] The earth's relative position to Jupiter moves in front of or behind the axis of Jupiter to the Sun, as the earth rotates around the Sun.

A recently analysed cuneiform tablet shows a table (Figure 11) of computed distances (synodic arc)*[580] and times (synodic time) for movement of Jupiter in relation to the earth. The table gives the exact location of Jupiter at particular times in relation to the zodiac and was calculated using a set of algorithms called Jupiter system A′ (Figure 12).[581] This system uses four discrete step functions

---

[576]   Britton, J.P., Walker, C. B. F. (1996), 'Astronomy and Astrology in Mesopotamia,' in *Astronomy before the Telescope,* ed. Walker, C. B. F., London: British Museum Press, p. 44.

[577]   The time a planet takes to rotate around its axis in relation to the surrounding stars.

[578]   Steele, J. M., Stephenson, F. R. (1997), 'Lunar Eclipse Times Predicted by the Babylonians,' *Journal for the History of Astronomy,* **xxviii** (2), pp. 119–131

[579]   Jupiter takes 11.86 Earth years to rotate around the Sun. Its orbital speed around the Sun is about half that of Earth.

[580]   Synodic arc and synodic time are displacement along the zodiac and the corresponding time.

[581]   Taken from Ossendrijver, M. (2015) *Ibid.,* System A′ is one of about ten known computational systems the Babylonians had for Jupiter.

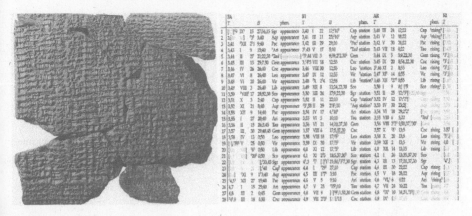

Figure 11: Tablet and translation: a computed table of movements of Jupiter. Source: Ossendrijver, M. (2015) 'Babylonian mathematical astronomy,' Researchgate at *https://www.researchgate.net/publication/278681968*.

Jupiter system A': parameters of the step function for the synodic arc

j	Zone	σ_j	r_j
1	9 ° Cnc – 9 ° Sco	30 °	1;7,30
2	9 °Sco – 2 °Cap	33;45 °	1;4
3	2 °Cap – 17 °Tau	36 °	0;56,15
4	17 °Tau – 9 °Cnc	33;45 °	0;53,20

Figure 12: Jupiter system A' parameters.

(see middle column: σ which is Jupiter's displacement along the zodiac) in calculating the position of Jupiter.

Two systems (now called Systems A and B by historians) were used by the Babylonians to calculate the position of the planets. System A (Figure 13), developed in 320 BC, was based on 'step functions' in which zones of constant amplitude vary discontinuously at their boundaries, and their variable quantity is expressed as a function of longitude.[582] System B which was developed later

---

[582] Britton, J. P., Walker, C. B. F. (1996) 'Astronomy and Astrology in Mesopotamia,' in *Astronomy before the Telescope*, ed. Walker, C.B.F., London: British Museum Press, p. 55.

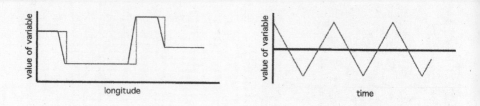

Figure 13: Modern depictions of a step function and a linear zigzag function used in Babylonian planetary movement prediction (L: system A; R: system B). Taken from Robson, E. (2008), p. 219.

(280 BC) was based on linear zigzag functions.[583] System B is characterised by the increasing or decreasing of a variable by a constant amount in successive intervals up to a fixed maximum or minimum (Figure 13. right side). Simple linear zigzag functions started to be developed as early as 2,000- 1,600 BC.[584] Patterns were also looked for in other apparently random atmospheric and terrestrial natural phenomena.

Babylonian Invention of Early Form of Calculus

It has recently been discovered[585] that Babylonians described a mathematical technique that was previously thought to have been discovered 1,400 years later and is an early form of calculus. The new analysis of previously undeciphered cuneiform tablets describe how the area of a trapezoidal figure (Figure 14) was used by the Babylonians to compute the position of Jupiter.

The tablet contains a previously unpublished trapezoid procedure for Jupiter's motion in the 60 days after its first appearance. The text describes the trapezoid but does not illustrate it. The figure described and now for the first time il-

---

[583] Hunger, H., Pingree, D.E. (1999), *Astral Sciences in Mesopotamia*. Leiden: Brill, p. 213.

[584] Robson, E. (2008), *Mathematics in Ancient Iraq: A Social History*, Princeton: Princeton University Press, p. 219.

[585] Ossendrijver, M. (2016), 'Ancient Babyonian astronomers calculated Jupiter's position from the area under a time-velocity graph,' *Science*, **351** (6272), pp. 482–484.

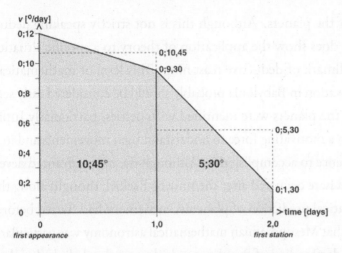

Figure 14: Time-velocity graph of Jupiter's motion. Daily displacement along the ecliptic *v*) between Jupiter's first appearance (day 0) and its first station (day 120) as a function of time according to scheme X. All numbers and axis labels are in sexagesimal place-value notation. The areas of the trapezoids, 10;45° and 5;30°, each represent Jupiter's total displacement during one interval of 60 days.

llustrated (Figure 14), documents the velocity of Jupiter against time in its motion across the ecliptic. The distance covered by Jupiter over 60 days is calculated as the area under the curve. This uses a method of integration that is a basic operation in calculus and is standard for modern astronomy, physics and mathematics.[586]

Summary: Babylonian Science

The use of step functions and the use of area under the curve integration to facilitate the study of the movement of Jupiter show that Babylonian mathematics was ahead of its time and ahead of what its neighbours were able to do. It also shows that they were able to rationally move to more complex mathematical methods in order to tackle the problem of the varying relative

---

[586] Cowen, M. (2016), 'Ancient Babylonians took first steps to calculus,' *Science*, **351** (6272), p. 435.

speeds of the planets. Although this is not strictly speaking deductive reasoning, it does show the application of theory to a specific situation, which is the hallmark of deductive reasoning. This level of mathematical enquiry and innovation in Babylonia probably should be considered early science. The fact that the planets were identified with deities, particularly Jupiter, must have been a motivating force to understand their movements and in a push towards science to accomplish this. Although the more advanced developments described here occurred after the time of Ezekiel, thought along the lines of mathematical prediction of planetary movement had started before his time. The fact that Mesopotamian mathematical astronomy was particularly applied to the understanding of the planets which were closely linked to their gods, is a parallel with the Chariot of YHWH, where divine and rational structures, are connected in keeping with 'divine science'.

Egyptian Science

Egyptian mathematics dates to about 3,000 BC with the start of the use of a decimal system and number symbols up to 100,000. The use of fractions is attested in the Rhind papyrus.[587] This papyrus was written after 1,800 BC, during the Hyksos domination of Egypt, but purportedly dates to an earlier time (2,000–1,800 BC).[588] Star calendars were developed at this time, but it was not till the New Kingdom (1,600–1,100 BC) that astronomy started. The achievements of Egyptian astronomy include: knowledge of stellar constellations - at least 43 constellations were familiar to the Egyptians in the 13th century BC; knowledge of planetary astronomy – five planets were known to the Egyptians; the retrograde motion of Mars was known; the revolution of

---

[587]   Peet, T.E. (1923), *The Rhind Mathematical Papyrus*, London: Hodder and Stoughton.
[588]   van der Waerden, B. L. (1975), *Science Awakening*, Leiden: Nordhoff International, p. 16.

Mercury and Venus around the Sun was known.[589] Astronomy may have been used in positioning the pyramids, (The pyramids are aligned very accurately, the eastern and western sides run almost due north and the southern and northern sides run almost due west.).[590,][591] The Egyptians also made a calendar of 360 days with twelve months of thirty days and five extra days of festivities added each year to make up to 365.

The Edwin Smith Surgical papyrus gives a rational approach to medicine and surgery in ancient Egypt. It tabulates 48 cases of injury detailing the type of injury, the examination of the patient, the diagnosis, prognosis and treatment. It also recognizes the influence of the location of brain injury on parts of the body affected by paralysis.[592] It recommends taking of the pulse and the use of olfactory and visual clues in the making of a diagnosis. The fact that the papyrus shows how to reach a diagnosis based on different clinical features appears to be evidence of deductive reasoning and therefore of scientific thought. The Egyptians were great innovators and produced many technological inventions; there is little evidence of formal use of hypothesis driven investigation, however. Ancient Greek science flourished starting around the time of Socrates death in 399 BC,[593] which was after the time of Ezekiel.

Pre-Science in Ezekiel

There is an indirect link between the Chariot Vision movements and Babylonian pre-science. As noted above Babylonian produced sophisticated pre-

---

[589]   Novakovic, B. (2008), 'Senenmut: An Ancient Egyptian Astronomer,' *Astro-ph.*, 8 Jan. 2008, *https://arxiv.org/pdf/0801.1331.pdf*.

[590]   Thurston, H. (2003), 'On the Orientation of Early Egyptian Pyramids,' *DIO*, **13** (1), p. 4.

[591]   This was probably for religious reasons. The Egyptian sun god Ra, was king of the Gods, patron of the Pharaoh and this orientation could be in deference to the path of the sun.

[592]   Breasted, J. H. (2010), *The Edwin Smith Surgical Papyrus: Hieroglyphic Transliteration, Translation and Commentary VI*, Chicago: University of Chicago Press, p. 12.

[593]   Van der Waerden, B. L. (1975), *Science Awakening*, Leiden: Nordhoff International, p. 149.

dictions of Jupiter's planetary movement using an early form of calculus. Babylonian planetary movement prediction system A was a step function (Figure 13) that appears similar to the right-angle lightning movements that the Chariot of YHWH makes. The Babylonian study of planets which were held to be divine has a counterpart with the Living Creatures' faces which are linked to the zodiac constellations, the four corners of the cosmos and the omnipotence of YHWH.

More definite evidence of scientific reasoning is seen in the visions of Ezekiel. Measurement is a hallmark of science, as it is the way in which an investigator, turns unknown lengths into standard units (for example, cubits). The numbers of standard units allow objective conclusions to be reached, either by allowing an accurate artistic rendition (as in the case of the Idealised Temple Vision), or because the numbers ('the data') can be interrogated statistically to give new objective conclusions. Conclusions that are contained within the data are reached by *deductive* scientific reasoning; conclusions that are inferred from, but not contained within the data are reached by *inductive* scientific reasoning. In the case of the Idealised Temple Vision, measurement allows an accurate scale drawing of the visualised temple to be made, to more easily appreciate its purpose. In this vision, the conclusion - the shape of the building, is contained within the measurements, so the process is similar to deductive reasoning.

An example of inductive reasoning is seen in the Vision of Water Flowing from the Temple: Ezekiel sees water flowing from below the right side of the threshold of the Temple (Ez 47:1-12), (Figure 16). The river flows toward the east and increases exponenttially in size as it flows. A man with a measuring stick (upper left corner of Figure 16) measures off four distances down the river, each of 1000 cubits (1000 cubits is about 450 metres). He tells Ezekiel to walk through the water at each of these points and he finds that at 1000 cubits it was ankle deep, at 2000 cubits it was knee-deep, at 3000 cubits it was up to the loins and at 4000 cubits it was too deep to cross and was deep enough to swim in. Ezekiel was asked to consider what this meant. The man with the measuring stick asks Ezekiel to make a conclusion that is not contained within the information given him, which is similar to inductive scientific reasoning.

The Wheel: Analogy for Science and Technology

It was suggested above (Lower Tier: Discussion) that, because of their concrete artificial construction, the wheels may be an analogy for measurement. It was also mentioned that measurement is a hallmark of science. It was also suggested that the wheels may be an analogy for inferior reason. Inferior reason would include what is within the remit of empirical science since inferior reason is a practical discipline dealing with 'the concrete' but includes issues of space and time,[594] and 'natural' science is a rational discipline that is orientated to the understanding of the changeable world. Since reason has the capacity to 'deliver a higher level of objective truth' the question then arises whether the wheels may not be an indication of truth that is revealed through science. This subsection will explore indications of the use of the wheel as a technological tool in Ezekiel as well as relevant mathematical properties of the wheel.

The wheel as a technological tool

Although simple in structure, the wheel is of central scientific importance as it has allowed many basic technological advances.[595] The critical structural features of the wheel that give it flexibility as a technological tool are:

---

[594] Chenu, M.-D. (1945), 'Ratio superior et inferior,' *Laval Théologique et Philosophique*, I (I), pp. 119–123, describes inferior reason as '*la connaissance des choses selon leurs causes propres et immédiates, à travers la dispersion des phénomènes et dans l'écoulement du temps: connaissance de la contingence, fragile mais très séduisante dans sa connaturalité avec notre esprit, et dans l'emprise de l'individuel concret*'.

[595] The wheel was also extremely important in ancient times since its discovery in Mesopotamia. Without the wheel for transport, pottery and in the war chariot (to name just three important early applications), society or civilisation might not have developed, at least as we know it.

1. Because of its round shape, when a wheel turns, its circumference effectively travels over a potentially infinite distance. This allows the turning wheel to travel unlimited distances, or if static to perform tasks (see below) that need to be repeated continuously.

2. Because the shape of a wheel expands out symmetrically from a central axis point, forces applied at its circumference act like a lever on the axis. The larger the wheel the smaller the force needed to turn it (torque). Also, when the wheel is turning, the angular velocity of a point on the circumference of the wheel increases as the wheel diameter increases.

3. When different sizes of wheels are connected as part of a geared device, in addition to the torque, the speed (by connecting different wheel sizes) and direction (by using toothed gears) of a power source can be altered. When used with a smooth axle and bearings, the wheel can reduce the power expended in transportation, by reducing the friction of dragging loads along the ground.

4. When static, a wheel acts as a rigid support between any two points connected across its diameter. In moving structures like a cart (or chariot), the wheel has a weight-bearing capacity, with the force of the load passing through its axis.

As a result of these features the wheel allows us to move loads, harness energy and transform forces. The principal modern technological applications of the wheel are:

- A movement platform for people and cargo (to replace walking, carrying or dragging) by reducing friction).

- Reducing power needed for heavy tasks: ship's wheel, pulley systems, gears.

- Power exchange with fluids such as water or air in hydro-electric or wind turbines, dynamos, jet engines, ship's propellers.

- Steam and combustion engines.

- Timepieces: watch gear systems.

- Making circular objects: potter's wheel, lathe.

In the ancient world, although there were fewer applications, civilisations which had the wheel had distinct advantages over those that didn't (e.g. potter's wheel, war chariots). Wheels were first used on carts in about 5,500 to 6,000 BC. The potter's wheel was first invented about 7,000 BC and was brought to Egypt in about 2,600 BC.[596] The chariot apparently originated in Mesopotamia around 3,000 BC,[597] but was first used in Egypt at the time of the Hyksos invasion in about 1,800 BC. Solid stone wheels had been used as pottery wheels before that time. Wheels that were used for measurement in Egypt were probably made of stone.[598]

The above-mentioned examples are only a small sample of all the applications of the wheel. The wheel's applications have increased over time. It is now a key component of thousands of modern inventions. The wheel can be said to be the principal material object that has effectively given man the technological ability to 'subdue the world' (Gn 1: 8).[599]

---

[596] Doherty, S. (2013), 'The origins and the use of the potter's wheel in Ancient Egypt,' PhD Thesis, Cardiff University.

[597] 'Chariot,' (2019), *Encyclopaedia Britannica* at: *https://www.britannica.com/technology/chariot* accessed on 25/9/2021

[598] Kushwah, A, 'Pi and the Great Pyramid of Giza,' at: *https://mathshrc.com/Journal/articles/a15/a15.html* accessed on 25/9/2021.

[599] Woodford C. (2017), 'Wheels,' at: http://www.explainthatstuff.com/howwheels-work.html Accessed 25/9/2021.

Spiritual character of the wheel

As mentioned above, a wheel also has a practical, weight-bearing capacity, with the force of the load passing through the axis and vertical diameter. The diameter is a fixed measurable dimension of a wheel. It therefore can represent the 'immanent' character of the wheel. Since the circumference of a wheel has no starting or finishing points for measurement, it can be considered to be of infinite dimension and can thus represent the 'transcendent' character of a wheel. Since the constant π (pi) is the ratio between a circle's circumference to its diameter (Figure 15), π can be seen as representing a balance between the immanent and transcendent aspects of a wheel or of a circle. Pi is an irrational number, which means it cannot be represented as a simple fraction. Irrational numbers cannot be represented as terminating or repeating decimals. Therefore, the digits of π go on forever in a seemingly random sequence, the first of which are 3.14159. The longest calculation performed in an attempt to find a finite value for π has reached 12 trillion digits.[600]

Technology based on the wheel in Ezekiel

There is one example of a technological tool based on the wheel in Ezekiel and this is balance scales. Ezekiel uses scales to divide his cut hair into three equal parts (Ez 5:1). Beam scales and equal-arm scales were used in ancient Egypt and have two arms extending from a central pivot and find the mass of an object by comparison with calibrated standard masses.[601] Scales have identical physics to the pivot and two horizontal spokes of a wheel (Figure 15). In addition, like a wheel, scales are considered a metaphor for justice (Ez 45:10 'You must use honest scales').

---

[600] Kondo, S., Yee, A. J., '12.1 trillion digits of Pi,' at *http://www.numberworld.org/misc_runs/pi-12t/* accessed 25/9/2021.

[601] Hebra A. (2014), *The Physics of Metrology*, New York: Springer, p. 103.

Figure 15: Spoked wheel with vertical diameter (*d*) the weight bearing or temporal dimension of the wheel) and the circumference of the wheel (*c*), the continuous (or spiritual) dimension. The endless nature of the rotation of a wheel invokes the infinite nature of the divine and lends itself to use as in a prayer wheel or a rosary. Wheels are also associated with the process of executing judgement (Ez 10:6) and Pr 20:26. The spiritual aspect of a wheel also links to the rotation of the moon around the earth and planets around the sun.

Conclusion: a science-based analogy of the chariot wheels?

Although the wheel is the most versatile technological tool it also has 'spiritual' attributes, being a perfect shape the circumference of which is not finitely describable by mathematics in relation to its radius. In the Chariot, a revelatory function is suggested by the measurable finite distance the wheels travel under the impetus of the spirit. In both the Idealised Temple Vision and the Vision of Water flowing from the Temple, measurement was able to produce a higher level of objective truth. The wheels could therefore be an indication of truth higher than simple measurement that is attainable through science-based methodology.

Comparison of Ezekiel's and Archimedes' experiments

Since Ezekiel's Vision of the Water Flowing from the Temple appears to have the structure of a scientific experiment, here it is compared with Archimedes'

Figure 16: Ezekiel's vision of water flowing from the temple. This image is on multiple internet bible study sites.

(288 BC) famous 'Eureka' experiment.[602] While he is relating the Idealised Temple Vision (Ez 40–48), Ezekiel sees water flowing from below the right side of the threshold of the temple (Ez 47:1–12) (Figure 16). Ezekiel is asked to measure the river at four 1000-cubit intervals and to make inferences about the associated increase in the level of water on his body at these four points.

Archimedes's experiment came about when he was asked by King Hieron II of Syracuse to tell if his crown was pure gold. Archimedes was puzzling on this when one day he was getting into his bathtub and noticed that as he got in, the water rose on the side of the tub in proportion to how far he was in the tub. He knew he had his answer. He intuitively recognised that the water in a tub is displaced upwards according to the volume of an object immersed in the water.

---

[602] Archimedes's experiment was first written about by Vitruvius in the first century BC and questions have been raised about the details of the story. Cf Biello, D. (2006), 'Fact or Fiction? Archimedes coined the Term "Eureka!" in the Bath,' *Scientific American*, At *https://www.scientificamerican.com/article/fact-or-fiction-archimede/* accessed 7 May 2022.

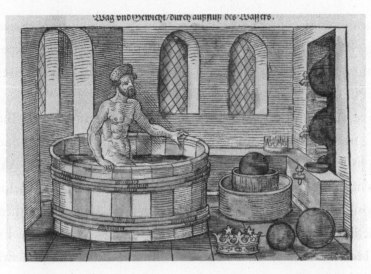

Figure 17: Archimedes' experiment (sixteenth-century illustration).

The similarities between these two 'experiments' are: (a) they both involve immersing a body in water; (b) the body is used as part of the measurement tool; (c) fixed objects are used as the scale of measurement (for Ezekiel: the measuring stick as the first scale and the height of his body as determined by key anatomical points as the second. For Archimedes: the inner surface of the tub as the first scale and the amount of his body immersed in water as the second); (d) they both have a 'eureka' conclusion: for Archimedes it was the law of displacement of water by a body; for Ezekiel it was the recognition of the unmeasurable nature of the water flowing from the Temple. The differences however are: (a) Ezekiel's is more sophisticated because not only is a standard distance carefully measured out setting the parameters of the experiment, but a second measurement is that of the height of water on his body. Archimedes just reckons the rising water level in the tub with his eye and compares this to how much of his body is immersed. (b) Whereas Archimedes measurement is a finite one as he measures his immersed body volume, Ezekiel's registers an apparent exponential increase in the river which is not measurable. Archimedes's observation can therefore be classed as a scientific experiment with a concrete result giving the scientific 'law of water displacement'. Ezekiel's experiment leaves him with an insight – the river is not behaving like a normal river and is the river of God's bounty which is ever increasing.

Two conclusions that can be drawn from this comparison are: (a) The Water Flowing from the Temple Vision is an experiment and this satisfies one of the criteria for the definition of science in this thesis given below.[603] Although there is no stated hypothesis, the statement by 'the man' at the conclusion of the vision (Ez 47:6) 'Son of man have you seen this?', suggests that there was a hypothesis and that this was 'that the water flowing from the Temple is a normal river'. However, the measurements show that it increases exponentially[604] in size and is therefore not a normal river and this has been demonstrated by the experiment; (b) Archimedes's experiment, too, had a hypothesis and an experiment. The hypothesis that 'water is displaced according to the volume of the object immersed' seems to have been thought of by Archimedes only as he immersed himself in the tub. He confirmed the hypothesis by noting that the water progressed up the side of the tub as he progressively lowered himself in.

Not only was Ezekiel surrounded by an active and advanced mathematics and early scientific environment but there are pointers that Ezekiel might have been aware of this: his recording of a sexagesimal unit of weight, a realisation of the importance of accurate weights and measures and the use of deductive and inductive reasoning tend to confirm this. Building on the rationality and frequent measurement in the Book, it does suggest that Ezekiel had access to early scientific ideas. The comparison of Ezekiel's Water Flowing from the Temple Vision with Archimedes experiment shows profound similarities and shows that both Archimedes immersion in the tub and Ezekiel's river

---

603 There must be both a hypothesis and an experiment, (see 'Hierarchies and Fuzzy Epistemology').

604 Here 'exponentially' means rapidly, not according to a specific mathematical exponential function.

measurement exercise fit the definition of early scientific experiments. Through measurement and rational structures, including a parallel of early science, Ezekiel's text shows that he is part of the precocious Babylonian pre-scientific movement and that these pre-scientific leanings also extend to the Chariot Vision.

Rationality in the Chariot Vision

Rationality is present at multiple levels in the Chariot Vision:

In the Upper Tier, there are two symbolic representations (above and below the loins of the 'likeness of a man') of both the rational and emotional aspects of YHWH's actions. The heart, represented by *hašmal,* is the centre of both rational (noetic) and emotional (affective) activities.[605] The *hašmal* therefore visually shows the need for the Israelites to re-integrate rationality into their decision making.*[606] The fire below the loins is also a rational symbol, since it can be interpreted as equating sexual and idolatrous 'promiscuity' and the need for purification (*kippēr*), through YHWH's personal intervention: 'when I forgive you' (*běkappēr-lak* בְּכַפְּרִי-לָךְ) (Ez 16:63).[607]

In the Middle Tier, it is argued that the movements of the living Creatures are reason-based and that 'moving straight ahead', may be a metaphor or analogy for 'right reason'. The interplay of the character-forces of the Living Creatures constitute a model for rational decision making (Figure 6) and the two main axes discerned are the cognitive (reason and insight/intuition) and affective/emotional. These two axes mirror the two main forces of decision making and give further detail to the rationality of how decision making in-

---

605 Launderville, D. F. (2007), *Spirit and Reason,* Waco: Baylor University Press, p. 348.

606 Electrum (the presumed meaning of *hašmal*) is a combination of two metals and could be an analogy of the joining together of rational and emotional forces.

607 This construct with YHWH as subject and humans as object is found only in three other places in the Bible: Nu 35:33; De 21:18 and Is 22:14: (Block, D.I. (1997), *The Book of Ezekiel Chapters 1–24,* Grand Rapids: Eerdmans, p. 520.

corporates emotional forces. Rationality is therefore found in the middle tier and the Living Creatures do support the presence of 'divine science': since the spirit dictates the movements of the Living Creatures, they could be said to reveal the divine in a rational way.

In the Lower Tier, the movements of the wheels are dictated by the Living Creatures' movements 'and when the living creatures went, the wheels went beside them' (Ez 1:19) and therefore are subject to rational forces that direct the Living Creatures. It is argued above that the wheels are a sign of measurement (metrology), and an analogy for inferior reason which includes scientific reasoning.[608]

In summary, evidence of rationality is apparent by chariot vision tier: illuminated and demonstrated through the chariot's Mesopotamian background; cosmological links; hierarchical levels; and potential revelatory roles of the tier structures. This is communicated through symbol, analogy or sign integrating the visual effect of the Chariot Vision and the text of Ezekiel. These rational connections are seen through cosmological, psychological, epistemological and metrological links, some of which are further discussed below.

Conclusion of the Chapter

Ezekiel uses measurement frequently and there are rational structures throughout the book. Instances of both superior and inferior reason are present with interactions: multiple measurements in two visions gave rise to higher levels of objective truth. Babylonians developed mathematical astronomy and their advanced mathematics was at a 'pre-scientific' level. Ezekiel's Vision of Water Flowing from the Temple may be classifiable as a scientific experiment. The wheels of the Chariot may represent science-based methodology.

---

[608] The science-fertile soil of Ezekiel suggests a scientific interpretation of the wheels of the Chariot Vision, particularly since there are grounds to consider the wheels an analogy for inferior reason. Within this context, therefore, the wheels of the chariot could be interpreted as a sign for revelation through empirical 'science'.

Chapter 7

Metaphor and Symbolic Reasoning

Overview of Chapter

This chapter looks at symbolism, and how it is built up in Ezekiel from metaphor. Metaphor as an indication of rationality is discussed in Mesopotamian texts and Ezekiel. Symbolism is reviewed both as a psychological process and as the principal method of human reasoning. Symbolism and symbolic thinking in Ezekiel are reviewed. A mathematical equivalent of symbolic reasoning, fuzzy logic, is discussed.

Ezekiel uses metaphor frequently and metaphors 'appear to be employed in an almost systematic fashion, and the arrangement of the metaphorical passages within the book appears to be a deliberate composition'.[609] It does appear that the whole book of Ezekiel is structured in an increasingly complex hierarchy of symbolism, going progressively from metaphor, to metaphoric narratives, sign actions, and the most detailed symbolism in his visions, of which the Chariot Vision could be said to be the most complex and mystical in the whole Bible. There appears therefore a rational hierarchy of symbolism. In the context of

---

[609] Biwul, J. (2013), *A Theological Examination of Symbolism in Ezekiel with Emphasis on the Shepherd Metaphor*, Carlisle: Langham, p. 4; Schöpflin, K. (2005), 'The Composition of Metaphorical Oracles within the book of Ezekiel,' *VT*, 55 (1), 102.

looking for evidence of rationality in Ezekiel, it is therefore important to look at metaphor and determine how symbolism contributes to reasoning and to Ezekiel's message.

Metaphors are potential Evidence of Rationality

Metaphors are symbolic comparisons between two different objects, one in the known 'real' world and the other in a 'possible' world, in order to bring out a specific feature. Metaphor highlights a chosen characteristic or feature often by comparing it to an extreme manifestation of that feature. A metaphor has a truth value as determined in a 'possible world' not in our real world.[610] It is therefore a contingent truth.*[611] Metaphors are rational, and can potentially have scientific value, for example, if they describe natural events. Biblical metaphors may represent cognitive structures that reveal deeper truths.[612] Metaphors used in the Bible therefore may have existed prior to their being written down. In this section I will look at metaphor use in Babylonia to determine the *Sitz im Leben* of Ezekielian metaphor. I will also review metaphors used in Ezekiel and try to determine if they have any specific rational role.

---

[610] Steinhart, E. (2001), *The Logic of Metaphor: Analogous Parts of Possible Worlds,* Berlin: Springer, p. 169.

[611] Metaphor is described by a special form of modal logic because metaphor forces a revision of belief. Modal logics can determine the difference between contingent and necessary truths. A contingent truth is true, but could have been false. A necessary truth is true but could not possibly have been false. Cf. Lietgeb, H., Segerberg, K. (2007), 'Dynamic Doxastic Logic: Why, How, and Where To?,' *Synthese,* 155 (2), pp. 167–190. When they achieve their goal, metaphors not only become part of the language but also develop a halo effect. That is, that every well-chosen metaphor can also be understood by those who have never heard it before. Also, for every metaphor there are an associated set of propositions that it makes sense to view in a different light. Cf. Segerberg, K. (2011), 'A Modal Logic of Metaphor,' *Studia Logica: An International Journal for Symbolic Logic,* 99 (1/3), pp. 337–347; p. 343.

[612] Tilford, N.E. (2017), *Sensing World, Sensing Wisdom: The Cognitive Foundation of Biblical Metaphors,* Atlanta: SBL Press.

Review of Babylonian Metaphor

Natural phenomena were often represented in Babylonian myth. Myth often incorporates these phenomena into imaginary entities to attempt to understand them.[613] This is particularly true of *omina*, which were predictive statements indicating the meaning of certain celestial events. Neo-Assyrian court astrologers regularly quoted these celestial *omina* and there is evidence that during the reign of Ashurbanipal in the seventh century, the king used astrological reports on a regular basis.[614] An omen is typically a conditional sentence, made up of a protase (that is a condition to be fulfilled) and an apodosis, (that is, the consequence if the stated condition is fulfilled). The protases of *omina* frequently relate to physical appearances of phenomena observed in a heavenly body. Figurative language was used for alterations seen in the Moon, Sun or Venus; for example, the Moon 'in distress' may relate to a lunar eclipse.[615] Part of the reason that *omina* focused particularly on the physical, natural world may have been that there was a motivation to understand natural phenomena, particularly those related to astrology. The Mesopotamian cuneiform writings that contain natural phenomena include *Enūma Anu Enlil*, the celestial omen series; *Šumma ālu* that deals with fauna; *Šumma izbu* about anomalous human births; *Alamdimmu* on anatomy; and parts of the *Ziqiqu* dreambook. In the celestial omen series *Enūma Anu Enlil*, a lunar eclipse is described as 'the moon god in mourning'.[616] Rochberg shows

---

[613]   This process would qualify for one of Wartofsky's pre-scientific modes of knowing. Cf. Wartofsky, M. W. (1968), *Conceptual Foundations of Scientific Thought*, New York: Collier-Macmillan, p. 61.

[614]   Koch-Westenholz, U. (1995), *Mesopotamian Astrology: An Introduction to Babylonian and Assyrian Celestial Divination*, Copenhagen: Carsten Niebuhr Institute of Ancient Near Eastern Studies, p. 52.

[615]   Isaiah uses 'the moon shall be confounded, and the sun ashamed' for the apocalyptic dimming of their light (Is 24:23).

[616]   Rochberg, F. (1996), 'Personifications and Metaphors in Babylonian Celestial Omina,' *Journal of the American Oriental Society*, **116** (3), pp. 475–485; p. 475.

convincingly that this figurative language is a metaphor. Since the name of the moon is indistinguishable from the name of the moon god, the figurative language could be relating to the moon or to the god. Rochberg argues that it is anthropomorphic metaphorical language and does not relate directly to the god, but it is an attempt to convey the appearance of something observed. Within the metaphor is a visual association that likens the observed appearance to a mental association that is elicited by it and is in fact abstract analogical thinking. It represents intellectual inquiry about nature which is consistent with what constitutes modern science. [617]

In conclusion, metaphor was in use in the royal courts as *omina*. Since metaphor is used in *omina* to describe physical phenomena, it may be that it was the principal context for intellectual inquiry about nature. Metaphor may have been used because Babylonians did not understand or have precise language to describe what they were seeing.[618] Metaphor has truth value and is a powerful tool for extending language[619] and has been thought to represent an equivalent to science.

Metaphor in Ezekiel

In Ezekiel, metaphors are often extended into metaphoric narratives, or a story that builds a picture with several metaphoric features supporting each other. Ezekiel tends to focus on a few basic metaphors and repeatedly modifies them.[620] His oracles are 'treasure troves for students of metaphor'.[621] Use

---

[617]   Rochberg, F. (1996), *Ibid.,* p. 484.

[618]   Segerberg, K. (2011), 'A Modal Logic of Metaphor,' *Studia Logica: An International Journal for Symbolic Logic,* **99** (1/3), pp. 337–347; p. 337.

[619]   Segerberg, K. (2011), *Ibid.,* p. 346.

[620]   Schöpflin, K. (2005), 'The Composition of Metaphorical Oracles within the Book of Ezekiel,' *VT,* **55** (1), pp. 101–120; p. 102.

[621]   Darr, K. P. (1994), 'Ezekiel among the Critics,' *CBR,* **2**, pp. 9–24.

of metaphor is also seen in Ez 16 and Ez 23, where the City of Jerusalem is portrayed as a metaphor of YHWH's wife by Ezekiel.[622]

Questions that arise about the use of metaphor in Ezekiel are: (a) why does Ezekiel use metaphor so frequently? (b) Is metaphor used only for rhetorical reasons or do the metaphors carry a moral message? (c) Are the metaphors evidence of rationality?

The metaphorical narratives in Ezekiel

Since the metaphorical narratives are extended metaphorical pictures on a single theme, they are probably the best place to start in trying to elucidate why Ezekiel uses metaphor. There are seven principal metaphorical narratives in Ezekiel:[623] The Shepherd Narrative (Ez 34); The Vine (Ez 15, 17, 19:10–14); The lioness and her cubs (Ez 19:1–9); Adultery (Ez 16, 23); Sinking of the Ship Tyre (Ez 27:1–36), King of Tyre (Ez 28:11–19); Cutting down of the Cedar – as Pharaoh (Ez 31:1-18); Capturing of the Water Monster – as Pharaoh (Ez 29:1–6, 32:1–16).

Some of these narratives have a single restated message, others like the Shepherd Narrative have multiple meanings. Within the single overall metaphor, the Shepherd Narrative is an indictment of failed leadership;[624] also a promise to restore the covenant with Israel, that the Lord will shepherd his people and also that the Messiah, the servant of YHWH will be of the house of David.[625] The metaphor conveys both a warning of YHWH's anger and judgment as well as a promise of restitution.

---

[622] Galambush, J. (1991), *Jerusalem in the Book of Ezekiel: The City as Yahweh's Wife*, Atlanta: Scholars' Press, p. 124.

[623] Durlesser, J. (2006), *The Metaphorical Narratives in the Book of Ezekiel*, Lewiston: Mellen, p. 9.

[624] Biwul, J. (2013), *A Theological Exmination of Symbolism in Ezekiel with Emphasis on the Shepherd Metaphor*, Carlisle: Langham, p. 275.

[625] Block, D. I. (1998), *The Book of Ezekiel Chapters 25–48*, Grand Rapids: Eerdmans, p. 308.

The vine is used metaphorically in 3 passages. The first (Ez 15) is that the vine is the residents of Jerusalem that will be destroyed for their faithlessness.[626] In the second (Ez 17:6) the vine is King Zedekiah who turns against God and is killed by Nebuchadnezzar. The third (Ez 19:10), is that the nation of Judah has become arrogant and is subjected to the Babylonians.[627]

In the metaphor of the Lioness and her Cubs (Ez 19;1–9), the lioness is likely a metaphor for Israel/Judah or the Davidic line and the cubs, the Kings. God shows that it is not enough for David's descendant to be on the throne, but the king must submit to the rule of God.[628]

The Adulterous Wife narrative (Ez 16:1-63) is a metaphor for the despising of the intense love of God and its self-inflicted perdition.[629] The same meaning can be applied to Ez 23. Devotion to any object or person other than God is considered adultery.

The narrative on the Ship Tyre (Ez 27) is a metaphor on 'pride goes before destruction' (Pr 16:18) as is that on the King of Tyre (Ez 28:11–19). Since Tyre's success was pursued in defiance of God, it was bound to fall.[630]

The narrative of the cedar tree represents the pride of Pharaoh, which draws the fury of God.[631] Narratives about a water monster (Ez 29:1-6, 32:1-16) are also metaphors for Pharaoh and through him, for pride. It warns that God is the Lord of history and will bring down the proud at the time he has set.[632]

The question why Ezekiel uses metaphor so frequently is partly answered by this quote from Golding:[633]

---

[626] Block, D. I. (1997), *The Book of Ezekiel Chapters 1–24*, Grand Rapids: Eerdmans, p. 459.

[627] Block, D. I. (1997), *Ibid.*, p. 609.

[628] Block, D.I. (1997) *Ibid.* p. 611.

[629] Block, D. I. (1997), *Ibid.*, p. 504.

[630] Block, D. I. (1998), *Ibid.*, p. 87.

[631] Block, D. I. (1998), *Ibid.*, p. 197.

[632] Block, D. I. (1998), *Ibid.*, p. 212.

[633] Golding, T. A. (2006), 'The Imagery of Shepherding in the Bible, Part 1,' *BS*, **163** (649), pp. 18–28; p. 20.

How does God, who is infinite beyond human experience and comprehension explain Himself and his relationship to people in ways they can understand?

The answer is through figurative and metaphorical language.

In Ezekiel, YHWH's message is uncompromising and direct and requires the use of symbol and metaphor to reach a hardened audience.[634] Ezekiel's writing needs to be infused with metaphor because the subject matter is complex and difficult to understand.[635]

The subject matter of the metaphors reviewed above shows they are not only used for their rhetorical impact but also to present a moral message. Bringing together the moral messages behind these oracles, there are three main themes: Putting other things before God brings his anger; pride brings humiliation; the despising of God's intense love brings self-inflicted perdition.

In terms of evidence of rationality, there is an underlying theme throughout the book of the divine Shepherding metaphor motif:[636] that human shepherds have failed to reverence YHWH and his name and also containing a promise of restitution. These metaphor narratives are therefore a rational vehicle for key messages of Ezekiel including: that YHWH loves deeply, that he expects nothing must come between his people and him and that pride distorts man and is not tolerated by YHWH.

Use of Discrete Metaphors in Ezekiel

Discrete metaphors are used frequently in Ezekiel but they must be distinguished from symbols.[637] Whereas a metaphor compares and contrasts two

---

[634] Block, D. I. (1997), *The Book of Ezekiel Chapters 1–24*, Grand Rapids: Eerdmans, p. 15.

[635] Duguid, I. M. (1994), *Ezekiel and the Leaders of Israel*, Leiden: Brill, p. 1.

[636] Biwul, J. (2013), *A Theological Examination of Symbolism in Ezekiel with Emphasis on the Shepherd Metaphor*, Carlisle: Langham, p. 4.

[637] Chisholm (Chisholm, R. B. (1998), *From Exegesis to Exposition*, Grand Rapids: Baker, p. 173) says 'A symbol represents something or someone by association, resemblance or con-

disparate images to bring out a crucial point they have in common, a symbol stands for a thing or person that is more complex and nuanced than itself. There is a resemblance between the way a symbol signifies and the way a metaphor communicates its truth through the tension between its literal and figurative image.[638] A metaphor can convey a symbolic idea, but a metaphor is the *process* of juxtaposing two things, rather than a symbol which points beyond itself to something else.[639] A sign also points beyond itself, but to something that has a transparent univocal meaning. Symbols in Ezekiel will be dealt with later in this section.

Discrete metaphors in Ezekiel can be grouped into several main topics:

1. As indications to Ezekiel on his course of action: Watchman for the House of Israel (Ez 3:16; 33:1); Ezekiel's silence - as a metaphor for initiation into divine service[640] (Ez 3:26); placing wood in Ezekiel's hand – Ezekiel himself becomes a metaphor for God (Ez 37:20).

2. To describe the people of Israel: as a harlot (Ez 6:9); a soothing aroma (Ez 20:41); a blossoming staff - wickedness (Ez 7:10); a grapevine consigned to fire (Ez 15:6); useless burned wood (Ez 15:5); bad shepherds – leaders exploiting people (Ez 34:7); a flock of sheep (Ez 36:38); dry bones – as spiritually dead (Ez 37:2); 'our bones are dried up' – despondency of exiles (Ez 37:11); piece of wood - one people (Ez 37: 19).

---

vention.' Ricoeur (Ricoeur, P. (1970), *Freud and Philosophy: an Essay on Interpretation,* New Haven: Yale University Press, p. 11) says that the term 'symbol' should be reserved for expressions that are plurivocal as opposed to 'sign' which has a univocal meaning. A symbol communicates a meaning about the thing or person signified, but there is an overt superficial meaning and a deeper concealed meaning as well. If the deeper meaning is absent, then it is called a sign. Cf: Launderville, D. F. (2007), *Spirit and Reason,* Waco: Baylor University Press, p. 6.

[638] Launderville, D. F. (2007), *Ibid.,* p. 76.

[639] Tillich, P. (2007), 'Religious Language as Symbolic' in *Philosophy of Religion: Selected Readings,* Peterson, M. *et al.* (eds), Oxford: Oxford University Press, pp. 435–41; p. 435.

[640] Block, D. I. (1997), *The Book of Ezekiel Chapters 1-24,* Grand Rapids: Eerdmans, p. 159.

3. To illustrate YHWH's emotions: I poured out my fury (Ez 36:18); I will open your graves - compassion and restoration of exiles[641] (Ez 37:12).

4. To describe Israel's enemies: Pharaoh as a reed crutch (Ez 29:6); pride (Ez 30:4); uncircumcised (Ez 32:21); a land that devours its people – those against God are his enemies[642] (Ez 36:13); Gog... I will turn you around – as a steed (Ez 38:4); storm cloud – for a large army (Ez 38:9); animal names – for nobility who are to be devoured (Ez 39: 18).

Other examples of specific uses of metaphors in Ezekiel are: Ez 36:17 Israel's defilement is compared to a woman in her menstrual period.[643] Here Ezekiel uses this as a metaphor to explain that the land and the sanctuary had been defiled and why YHWH departed; metaphor of the cauldron, Ez 11:2:[644] YHWH turns the metaphor into an indictment of the men of Jerusalem who have apostatised; as an indication to Ezekiel, how his actions should be (Ez 3:7), 'the house of Israel are of a hard forehead and a stubborn heart'; to the people of Israel, Ez 36:26 'I will take out of your flesh the heart of stone and give you are heart of flesh',*[645] (The phrase that follows 'and I will put my spirit within you' is not however a metaphor as *rûaḥ* is the spirit that animates the body[646]). As discussed below this is a key metaphor in the book of Ezekiel as the stone relates to a lack of the spirit and a defect of the operation of reason. This results in a loss of integrative rationality and symbolic thinking.[647]

In summary metaphor in Ezekiel is particularly focused on the Israelites: their spiritual state in relation to YHWH, and also on their enemies. Metaphor

---

[641] Block, D. I. (1998), *The Book of Ezekiel Chapters 25–48*, Grand Rapids: Eerdmans, p. 387.

[642] Block D. I. (1998), *Ibid.*, p. 336.

[643] Launderville, D. F. (2007), *Spirit and Reason*, Waco: Baylor University Press, p. 97.

[644] Launderville, D. F. (2007), *Ibid.*, p. 118.

[645] This is reminiscent of De 30:6 'YHWH your God will circumcise your heart and the heart of your descendants to love your God with your whole heart.'

[646] Launderville, D. F. (2007,) *Spirit and Reason*, Waco: Baylor University Press, p. 352.

[647] Launderville, D. F. (2007), *Ibid.*, p. 348.

is used in YHWH's communication with Ezekiel to emphasize specific points and also in speaking directly to the Israelites. In most cases, it is to warn about their dire situation in relation to YHWH, or to show how Israel's enemies' pride will be their downfall. Typical of metaphors, their use is almost always to illustrate a truth that may not be otherwise appreciated or that people have become inured to. Ezekiel uses metaphor in a systematic premeditated way. Many of the metaphors convey a symbolic idea, for example, in the metaphor of the cauldron, (Ez 11:2) the public officials use the metaphor of meat in a cauldron to show they are protected. YHWH (Ez 11:11) tells them they will not be protected by the walls of Jerusalem as in their metaphor and turns the pot into a symbol of their apostasy. Some of the metaphors act as a sign, for example in Ez 36:18, YHWH's fury is compared to molten metal.[648] In conclusion, use of metaphor in Ezekiel is rational as it is frequently an attempt to re-establish right reason in those who have become hardened to the action of the spirit. The following section on the metaphor of 'heart of stone' further explores this and the embedding of rationality within the motivation for using metaphor.

Specific role of metaphor in Ezekiel: the heart

The most central and powerful metaphor in Ezekiel is that of the heart of stone 'I will remove the heart of stone from your body and I will give you a heart of flesh': *wahăsirotî lēb hā'eben mibbĕśarām wĕnatattî lāhem lēb bāśār* (והסרתי לב האבן מבשרם ונתתי להם לב בשר) which is found in two verses; (Ez 11:19; 36:26). These two verses are virtual repetitions of each other. The first mention that the Israelites have a heart of stone comes just after the Vision of the Executioners, when the city of Jerusalem has been devastated and the Temple defiled because of the idolatry of the Israelites. Ez 11:14–21 is a powerful statement of YHWH's faithfulness to his covenant and that despite their ongoing sinfulness, the Lord will renew Israel spiritually.

---

[648] Block, D. I. (1998), *The Book of Ezekiel Chapters 25–48*, Grand Rapids: Eerdmans, p. 346.

Comparing their hearts to stone, is a metaphor meant to give the Israelites insight into the root of the problem that affects them. Their hearts were no longer receptive to his message and had become lifeless. This was not just a metaphor it was a diagnosis, and it was accompanied by a cure, a heart 'transplant': 'and I will give you a heart of flesh' (Ez 36:26). YHWH will purify Israel from their defilements (Ez 20:25): 'Let the sword come down...that their hearts may melt' (Ez 24:14–15), and YHWH will sprinkle them with clean water (Ez 36:25). He will cause the Israelites to be obedient to him and he will renew his covenant with his people.

Block sees a Jeremianic influence here because of a similarity with Je 31:33: 'I will put my law within them and I will write it upon their hearts' *natattî et-tôrōtay běqirbam wě'al-libbām ektăbbenă* (נתתי את תורתי בקרבם ועל לבם אכתבנה) which is similar to Ez 36:26 'and I will put a new spiritwithin them...' *wě-rûaḥ ḥădāšâ 'ettēn běqirběkem* (ורוח חדשה אתן בקרבכם).[649, 650]

A further metaphorical link with the heart is that the comparison of the Israelites' heart to stone, comes just after a statement of a new exodus for the Israelites (Ez 36:24): YHWH will reunite them with their land, fulfilling his patron responsibilities even as understood by other nations.

> I will take you from the nations,
> I will gather you from all the lands,
> I will bring you to your own land.

Ezekiel's metaphor of stone hearts gives further insight into the promise of De 30:6-8, 'Then YHWH your God will circumcise your heart and the heart of your descendants to love YHWH your God with your whole heart and your whole being in order that you may live...[651]

---

[649] Block, D. I. (1998), *The Book of Ezekiel Chapters 25–48*, Grand Rapids: Eerdmans, p. 356.

[650] The second phrase of Je 31:33 'and I will write it upon their hearts' links with Ezekiel's use of *lûaḥ* (a writing tablet) for the boards of the Ship Tyre, portraying its self-obsessed beauty (Ez 27:5), see Block, D. I.(1998), *Ibid.*, p. 59.

[651] Block, D. I. (1998), *Ibid.*, p. 354.

The heart of stone metaphor is key because the heart is the place where truth is discerned, decisions are made and the place from which deception issues.[652] Metaphor is concerned with transmitting a truth and so it is a mechanism of sensitizing the heart to truth. The heart is the place where cognition (noesis) and emotion (affection) are balanced,[653] and subconscious intuitive and affective forces affect our appreciation of truth. If the heart is hardened or desensitized to the truth it must be re-sensitized, and metaphor is one of the mechanisms by which Ezekiel does this. Unlike *omina* in Babylonia, Ezekiel was little concerned with external natural phenomena. The metaphor of the heart shows however that his metaphor impacted on key internal psychological decision-making mechanisms in a rational and scientific way. Balancing of cognition and emotion in the heart is embedded in the interaction of the character-forces of the Living Creatures and the metaphor of the heart can also be discerned in the 'likeness of a man' in the Chariot Vision. The above shows how central metaphor is to Ezekiel in that it is used to animate the central themes of YHWH's displeasure as well as his faithfulness to his covenant.

Science comes of age: From pre-science to deductive reasoning

One of the issues in the development of 'pre-science' to academic science, is to determine what the steps were from pure observation to objective science and at what stage it can be considered that science was taking place. In her review of Babylonian *omina*, Rochberg[654] has suggested that metaphoric figurative language is evidence of abstract analogical thinking. It represented intellectual inquiry about nature. However, there are different levels of intellectual inquiry starting from an observation and asking oneself what is observed. Formulating

---

[652] Launderville, D. F. (2007), *Spirit and Reason*, Waco: Baylor University Press, p. 232.

[653] Cf. Launderville, D. F. (2007), *Ibid.*, p. 45.

[654] Rochberg, F. (1999), 'Empiricism in Babylonian Omen Texts and the Classification of Mesopotamian Divination as Science,' *Journal of the American Oriental Society*, **119** (4), pp. 559–569; p. 569.

a metaphor about a phenomenon is maybe a step further, as the observer is beginning to compare the phenomenon to what elicits a similar mental picture. Rochberg notes that the Babylonian *omina* contain both empiricism and systematisation of knowledge and therefore deserve to be classified as science. The empirical part of the omen may be the observation made about the eclipse of the moon for example.[655] The systematisation of knowledge is the metaphorical description which can also be a 'theoretization' or a theoretical description.

Symbolic Reasoning

In Babylonian *omina*, it is likely that analogical thinking took place when an observed appearance was likened to a mental image. The question is whether this abstract analogical thinking was 'scientific'?[656] In this section we will look further at how humans use symbols in reasoning logically and arithmetically and how this 'symbolic' reasoning relates to science. This is important in the context of trying to establish at what point scientific thought started in Babylonia and also because of Ezekiel's frequent use of symbolism.

Symbolism as a mental process

Lakoff and Johnson put forward a conceptual metaphor theory that metaphors represent the fundamental cognitive mechanism by which humans normally think and are not simply rhetorical devices. Metaphors link an established mental understanding of an object to a novel understanding of the object. Mental processing however uses symbols to reason with[657] rather than using univocally

---

655 Rochberg, F. (1999), *Ibid.*, p. 559.

656 Rochberg, F. (1996), 'Personifications and Metaphors in Babylonian Celestial Omina,' *Journal of the American Oriental Society*, 116 (3), pp. 475–485; p. 484.

657 Landy, D., Allen, C., Zednik, C. (2014), 'A perceptual account of symbolic reasoning,' *Frontiers in Psychology*, 5, p. 1.

defined objects. Symbols not only point beyond themselves to something else, but also participate in the reality to which they point.[658] Objects in our brain are therefore manipulated as information structures and a metaphor seeks to link an existing structure to a novel one. [659] Symbolic reasoning allows us to more easily have an intuitive grasp of things around us.[660] This intuitive apprehension of reality is a link between our rationality and our emotion. For science, precision in value is essential. For emotion, precision in meaning is important. For symbolic reasoning, both are important.[661]

Psychological Background to Symbolic Reasoning

A traditionalist view of how people reason arithmetically and logically is that we use 'amodal'*[662] inner symbols to represent abstract data such as mathematics and logical propositions, and to complete imprecise perceptions. These symbols are manipulated according to internally represented mathematical and logical rules.[663] Problems can be mentally approached either by syntactic or

---

[658] Launderville, D. F. (2007), *Spirit and Reason*, Waco: Baylor University Press, p. 7.

[659] Lakoff, G., Johnson, M. (1980), *Metaphors We Live By*, Chicago: Chicago University Press, p. 258: 'metaphors link source domain neural information structures (parameterizations) to target domain parameterizations'. For example, the metaphor 'argument is war' is one of the metaphors we live by as it structures the actions we perform, when we argue.

[660] Launderville D. F. (2007), *Ibid.*, p. 50; This links to the concept of symbolic reasoning in the psychological theory of 'Affordances' put forward by Gibson in 1979 which states that 'affordances', or clues in the environment that indicate possibilities for human action, are perceived in a direct, immediate way with no formal sensory processing. Cf. Gibson, J. J. (1986), *The Ecological Approach to Visual Perception*, Hillsdale: Lawrence Erlbaum, p. 127.

[661] Cassirer says that humans are more accurately described as symbolic animals than rational: Cassirer, E. (1944), *Essay on Man*, New Haven: Yale University Press, p. 26: This is because symbolic activity brings together intellectual, volitional and emotional activity and does not just determine what is real on the basis of discursive reason.

[662] An amodal perception is the perception of the whole of a physical structure when only parts of it affect the sensory receptors.

[663] Landy, D., Allen, C. Zednik, C. (2014) *Ibid.*

semantic groupings. There may be personal emphasis by individuals to one of these ways.

Two main theories of how symbolic reasoning occurs are computation and semantic processing. The idea behind computation[664] is that reasoning is performed by manipulation of symbols according to syntactic rules. The computational view is that reasoning takes place in a central system that is isolated from sensorimotor input.[665] It has been suggested that when a sentence is read, the brain registers the meaning of the characters drawn on a page and converts them into a symbolic auxiliary language called 'Semantic Markerese'.[666] The mental symbols and expressions of this language are operated on by syntactic rules that contain mathematical and logical fundamentals, and these rules are thought to be structures that control the use of internal symbols. There appear to be cognitive mechanisms (like mathematics) that cut across cognitive domains.[667] For example, we often make representations of things on the basis of incomplete information coming into one cognitive domain (e.g. language) but we may have information in other domains (e.g. visuo-spatial) that bear on this. We therefore need to correct these representations according to what memory holds in these other domains, and this mechanism of correction, called 'fixation of perceptual belief', cuts across domains.[668] This is also true of language use, which needs simultaneous access to multiple domains. Once a new perception has been integrated across domains, there still remains the need for veridicality*[669] of the perception in order to use this information to determine

---

[664] Anderson, J. R. (2007), *How Can the Human Mind Occur in the Physical Universe?*, New York: Oxford University Press, p. 31.

[665] Pylyshyn, Z. (1999), 'Is vision continuous with cognition? The case for cognitive impenetrability of visual perception,' *Behav. Brain Sci.*, **22** (3), pp. 341–423.

[666] Lewis, D. K. (1970), 'General semantics,' *Synthese*, **22** (1/2), pp. 18–67.

[667] Cognitive domains describe the main areas of brain functioning for example: language, visuospatial, memory.

[668] Fodor, J. A. (1983), *The Modularity of Mind*, Cambridge, Mass.: Bradford, p. 102.

[669] Veridicality is the semantic or grammatical assertion of the truth. Judging the truth is through the action of the heart. Ezekiel gains the ability to 'see to the heart of the matter'

how we ought to act. The mechanism to do this must also have access to all domains. Once a symbolic solution to a sensory input has been computed in the brain, it is converted back into written or spoken language.[670]

A semantic approach to symbolic reasoning is called Semantic Processing. In this approach, symbolic reasoning occurs through systems that interpret and represent mathe-matical and logical rules. Semantic processing theories include mental models,[671] and sensorimotor simulations of objects or scenes.[672] These theories hold that symbolic reasoning does not relate to syntactical rules but to the recognition of our understanding of the task in mathematical or logical terms. In mathematical reasoning, there is growing evidence we may have an internal spatial representation of numerical magnitude and that when we calculate mentally, our attention moves along this internal spatial representation.

A recent research study to test this, first distracted subjects' attention by running stripes in front of their eyes, causing rapid jerks of the eyes (as when looking out of a train). At the same time, the subjects were asked to perform mental arithmetic, performing addition or subtraction of two-digit numbers spoken to them. The stripe distraction caused: (a) an underestimation of addition and an overestimation of subtraction, (b) a downward shift of eye gaze when calculating subtractions and an upward shift of eye gaze when calculating additions. The authors concluded that spatial processing is pervasive in mental arithmetic.[673] This study shows that when mentally manipulating arbitrary

---

or a capacity to perceive the truth through his visions. Launderville, D. F. (2007), *Spirit and Reason,* Waco: Baylor University Press, p. 3.

[670] Landy, D. (2014), 'A perceptual account of symbolic reasoning,' *Frontiers in Psychology,* **5,** p. 2.

[671] Johnson-Laird, P. N. (1992), 'Propositional reasoning by model,' *Psychol. Rev.,* **99** (3), pp. 418–493.

[672] Barsalou, L. W. (1999), 'Perceptual symbol systems,' *Behav. Brain Sci.,* **22** (4), pp. 577–660.

[673] Blini, E. (2019), 'Spatial grounding of symbolic arithmetic: an investigation with optokinetic stimulation,' *Psychol. Res.,* **83** (1), pp. 64–83.

symbolic tokens, our capacity for symbolic reasoning does depend on the ability to use abstract mathematical and logical rules.[674]

An important factor underlying semantic reasoning is truth. Fodor stresses that semantic language has to treat the conditions under which the symbol would be true.[675] 'Semantics with no treatment of truth is not semantics'.[676] Since science is concerned with the truth of what underlies natural phenomena the subjection of semantic reasoning to truth is at the core of science, and also of the pre-scientific value of metaphor. Ezekiel shows that humans have a capacity to interpret symbols truthfully by calling upon the capacities of reason and spirit.[677]

Computation and Semantic Processing both are 'translational' explanations.[678] These hold that the processes of perception (the sensory organs, vision and touch) are only mediators of information from the outside world but are not involved in reasoning. They assume that any transformation in semantic structure occurs internally by Markerese expression, mental modelling or metaphor formation. There is a newer view, however called 'constitutive' in which the external sensorimotor processes are actually part of the process that simplifies the semantic structure of the problem rather than being a passive conduit of information. The 'Cyborg' view[679] of simple reasoning is an example of a constitutive formulation. It holds that notations made by a person externally can actually 'scaffold' the internal processes involved in symbolic reasoning. These notations allow storing, inspection, deletion and manipulation of in-

---

[674] Landy, D. (2014), 'A perceptual account of symbolic reasoning,' *Frontiers in Psychology*, 5, p. 1.

[675] Fodor, J. A. (1975), *The Language of Thought*, New York: Crowell, p. 120.

[676] Lewis, D. K. (1970), 'General semantics,' *Synthese*, **22** (1/2); p. 18.

[677] Launderville, D. F. (2007), *Spirit and Reason*, Waco: Baylor University Press, p. 94.

[678] Landy, D. (2014), *Ibid.*, **5**, p. 3.

[679] Clark, A. (2003), *Natural Born Cyborgs*, Oxford: Oxford University Press.

formation in way that facilitates the execution of symbolic reasoning tasks.[680] This helps the speed and accuracy and complexity of symbolic reasoning tasks. A well-known example of this is the notation on paper of 'carrying' of a digit during a complex multiplication.

A newer theory of symbolic reasoning that builds on the cyborg view is Perceptual Manipulations Theory. This suggests that perceptual processing is important for central symbolic reasoning. Mathematical and logical equations may be centrally learned or encoded in a perceptually rich format where details of spacing are important. The visual system is particularly responsive to dynamic movement, so this too is important. Learning how to perceptually and physically utilise notations (for example written cues during calculating large numbers) in accordance with their mathematical meanings, is critical to the capacity for reasoning.[681,*682]

In summary, this section has reviewed the symbolic processes of reasoning. It shows that the critical central features of symbolic reasoning are that (a) it uses incomplete or amodal symbols, (b) it is different from purely rational reasoning which is slower, more precise and does not involve emotional input,

---

680   Landy, D. (2014), 'A perceptual account of symbolic reasoning,' *Frontiers in Psychology*, 5, p. 2.

681   Landy, D. (2014), *Ibid.*, p. 7.

682   In connection with Babylonian mathematics, the suggested use of finger creases to count up to 60 in the sexagesimal system would be an early use of 'scaffolding' to assist central symbolic reasoning (see section 'Babylonia – Origin of Mathematics'). The cuneiform system of number writing (Figure 8) placed much importance on spacing for visual assessment of numerical value. In addition, the Babylonians made extensive use of pre-calculated tables for difficult calculations such as squares and square roots. These are examples of facilitation of mathematical symbolic reasoning by an external notation process and are very much in line with the Perceptual Manipulations Theory and its stated importance of the use of notations for mathematical reasoning. The foregoing gives insight into how the development of the cuneiform script and sexagesimal numbering system, which were powerful, efficient and easily learned systems of formal mathematical and logical notation, facilitated the development of Babylonian mathematical expertise. Cf. Dantzig, T. (2007), *Number: The Language of Science*, New York: Penguin (Plume), p. 21.

c) it can give an intuitive grasp of the reality of what is observed, d) It involves a link between our rationality and emotion or thinking and feeling, e) symbolic thinking is polyvocal and nuanced as opposed to purely rational thinking which is univocal and concrete.[683]

Symbolism and symbolic thinking in Ezekiel

In addition to frequent use of metaphor, Ezekiel uses symbols frequently. He uses so many symbols in his oracles that they 'seem to become the rule with Ezekiel'.[684] There are at least fourteen 'sign-acts'. These sometimes contain metaphors, but are more visual performances that have a specific symbolic message:[685]

1. Eating the scroll (Ez 2:8)

2. Ezekiel's dumbness (Ez 3:26)

3. Clay model of Jerusalem (Ez 4:1)

4. Lying on his left and right (Ez 4:4)

5. Rationing his food and water (Ez 4:9)

6. Shaving with a sharp sword (Ez 5:1)

7. Clapping his hands and stamping his feet (Ez 6:11) and bringing a sword down (Ez 21:14)

8. Packing and digging a hole in the wall and exiting (Ez 12:3)

9. Eating with trembling and fearfulness (Ez 12:18)

---

[683] Launderville, D. F. (2007), *Spirit and Reason*, Waco: Baylor University Press, p. 7.

[684] Soggin, J. A. (1980), *Introduction to the Old Testament*, London: SCM Press, p. 303.

[685] Biwul, J. (2013), *A Theological Examination of Symbolism in Ezekiel with Emphasis on the Shepherd Metaphor*, Carlisle: Langham, p. 37.

10. Groaning instead of crying (Ez 21:6)

11. Setting up a road sign (Ez 21:19)

12. Not mourning for his wife (Ez 24:15)

13. Resuscitation of an army from bones (Ez 37:1)

14. The two sticks (Ez 37:15)

The importance of these sign-acts relates to his audience, to the events that were going to happen, that YHWH wanted to warn the Israelites about. Of all the prophets, Ezekiel was probably the most colourful in his use of symbolism.[686] Ezekiel probably turned to symbolic actions rather than just declarations, because his audience were hard of heart and were hostile to his message. A living symbolism is put forward by Launderville, who says that Ezekiel fosters a type of symbolic thinking that made the Israelites living symbols of God. However, there was a fundamental flaw in the way that the Israelites perceived reality.[687] He therefore shows that a central task of Ezekiel was to restore the Israelites as living symbols of God through the interaction of reason and spirit. (see 'The Interaction of Reason and Spirit').

Symbolic reasoning, fuzzy systems and fuzzy logic

Fuzzy systems can be considered the scientific/mathematical counterpart of symbolic reasoning. Human perceptions and language are by nature imprecise. Zadeh*[688] was struck by how the brain can compute imprecise or incomplete

---

[686] Biwul, J. (2013), *A Theological Examination of Symbolism in Ezekiel with Emphasis on the Shepherd Metaphor*, Carlisle: Langham, p. 39.

[687] Launderville, D. F. (2007), *Spirit and Reason*, Waco: Baylor University Press, p. 348.

[688] Fuzzy logic was first described by L. Zadeh. who described a way to compute perceptions: Cf. Zadeh, L. A. (2002), 'From Computing with Numbers to Computing with Words – From Manipulation of Measurements to Manipulation of Perceptions,' *Int. J. Appl. Math. Comput. Sci.*, **12** (3), pp. 307–324.

perceptions and can manipulate these in a way that is different to a computer. He invented fuzzy logic, and fuzzy sets and systems, which mimic the way the brain normally manipulates perceptions. Fuzzy systems can 'compute' perceptions and language by representing them as multiple 'fuzzy sets' using 'Generalized Constraint Language' (GCL)*[689], GCL fuzzy sets are then 'defuzzified' back to 'crisp sets' that are mathematically manipulable by conventional computers. 'Defuzzification' applies a measurable structure to an uncertain observation or language. Fuzzy logic essentially tries to replicate the normal mental processes of utilizing imprecise perceptions, so it is a mathematical counterpart of 'symbolic reasoning' described by psychologists. The abundance of symbolism in Ezekiel suggests that fuzzy logic may have an application within Ezekiel and this will be explored further below.

Conclusion of the chapter

Ezekiel uses metaphor and symbolism as a rational vehicle for key messages. A crucial communication is that YHWH will replace the Israelites' hearts of stone with hearts of flesh making them into living symbols. The balancing of cognition and emotion takes place in the heart and is illustrated in the interaction of character forces of the Living Creatures. The brain preferentially uses symbolic reasoning as it enables an intuitive apprehension of reality that links cognition and emotion. Ezekiel used metaphor and symbolism to revive the Israelites flawed perception of reality.

---

[689] GCL uses multiple 'constraints' which describe the state of knowledge of an imprecise idea or word. GCL is similar to a symbolic auxiliary language like 'Semantic Markerese' mentioned above, in terms of the brain mechanism it attempts to replicate.

Chapter 8

Spirit in Ezekiel and the Chariot

Overview of Chapter

The meaning of the word *rûaḥ* in Ezekiel is explored and categorised. Actions of the spirit are looked for both in the text of Ezekiel and within the Chariot Vision.

Spirit *rûaḥ*

The three tiers of the Chariot Vision are linked to the spirit. In Chapter 3 (Middle Tier of Chariot Vision) it was shown that the Living Creatures 'moving straight ahead' without turning is an analogy for decision making, guided by the spirit. In Chapter 4 (Lower Tier of Chariot Vision) the wheels are also shown to be subject to this spirit. In Chapter 5, (Upper Tier of Chariot Vision), the 'likeness of a man' (Ez 1:26), seems to have a close association with the Spirit: in Ez 8:3, a man who has the same description as the 'likeness of a man' takes Ezekiel by a lock of hair, and 'the Spirit lifted me up'.

The spirit is therefore essential to the understanding of the Chariot Vision. For this reason, the role of the spirit in Ezekiel and particularly in the Chariot Vision will be examined in this chapter. Block says that Ezekiel can be described as the prophet of the spirit.[690] The word *rûaḥ* (רוּחַ), 'spirit', is used more in Ezekiel than in any of the other prophets and the spirit is a unifying theme across the book of Ezekiel. It links the Chariot Vision and the Vision of the Dry Bones as the spirit animates both the Living Creatures and the bones: the four Living Creatures of the Chariot of YHWH are animated and move by the spirit (Ez 1:20); Ezekiel is told to tell the bones that YHWH will put breath into them, and he does this by saying to the spirit, 'come from the four winds, O spirit and breathe upon these slain that they may live' (Ez 37:9). Launderville says that the spirit is 'the invisible reality that brought order and dynamism into the world described' in Ezekiel.[691]

The word *rûaḥ* is mostly used to mean an 'agency of animation' in Ezekiel.[692] The precise meaning of the word *rûaḥ* in Ezekiel however depends on its context and it can vary from breath/wind/life-spirit to an animating spirit, and even specifically to God's own Spirit (Ez 37:14).[693]

Block has outlined eight different meanings of *rûaḥ* in Ezekiel:[694] (1) Wind (a natural phenomenon) (Ez 37:9); (2) 'Direction' (of the winds) (Ez 5:2); (3) Compass point (Ez 42:16); (4) Agency of Conveyance- by which Ezekiel was lifted up and enters a trance (Ez 3:12): this is also described as the 'hand of YHWH' (Ez 3:14);[695] (5) Agency of animation: animating the Living Creatures

---

[690] Block, D. I. (2013), *By the River Chebar: History, Literary, and Theological Studies in the Book of Ezekiel*, Eugene: Cascade, p. 142.

[691] Launderville, D. (2004), 'Ezekiel's Throne-Chariot Vision: Spiritualizing the Model of Divine Royal Rule,' *Catholic Biblical Quarterly*, **66** (3), pp. 361–377; p. 375.

[692] Block, D. I. (2013), *Ibid.*, p. 150.

[693] Boadt, L. (2003), 'The Dramatic Structure of Ezekiel 37, 1–4,' in *Palabra, Prodigio, Poesia: In Memoriam P. Luis Alonso Schöckel, S. J.*, Analecta Biblica Dissertationes 151, Rome: Editrice Pontificio Istituto Biblico, 191–205; p. 203.

[694] Block, D. I. (1989), 'The Prophet of the Spirit: the Use of RWH in the Book of Ezekiel,' *JETS*, **32** (1), pp. 27–49.

[695] The animating force of the spirit is particularly seen in the expressions: 'the spirit of YHWH fell upon me' (Ez 11:5) or 'the hand of YHWH came upon me' (Ez 37:1; 40:1) when Ezekiel enters into the Vision of the Dry Bones and also the Idealised Temple Vision. There is no evidence to suggest that Ezekiel went on a literal journey in these

and the wheels (Ez 1:20). Block says that this can be attributed only to the presence of the life-giving spirit of God.[696] This is the most frequent way in which *rûaḥ* is used; (6) 'Agency of prophetic inspiration', for example Ez 2:2 'the spirit entered me' (it is unclear how this differs from the animating agency manifestation of the spirit however); (7) As 'mind' or 'heart' which reflects a common usage in the Old Testament (Ez 3:14);[697] (8) Divine ownership (Ez 39:29).

In looking more specifically at the manifestations of *rûaḥ* in the Chariot Vision, it is a better fit to use four main groups, categorised by an increasing manifestation of the divine Spirit of YHWH.

Using this classification, the different instances of *rûaḥ* are as follows:

1. At the lowest is the animation of natural phenomena (wind or water or fire) by the spirit: this incudes use of *rûaḥ* to denote wind or use of wind direction as a compass notification (Ez 42:16). Wind from the east (Ez 19:12) is used as a divine retributive force however (Ez 19:26), and all four winds together (Ez 43:9) are a spiritual force.

2. The second is the animation of beings with breath, spirit or mind by rûaḥ. For example: 'Every spirit shall be faint', *wakihătāh kāl rûaḥ* (וְכָהֲתָה כָל־רוּחַ) (Ez 21:12), or the breath that brings life or function to the human body: 'Thus says the Lord YHWH to these bones: Behold, I will cause breath to enter you, and you shall live' (Ez 37:5).

---

instances and more likely they represent trance-like prophetic experiences, during which he has vivid visions. (Block, D. I. (1998), *The Book of Ezekiel Chapters 25–48*, Grand Rapids: Eerdmans, p.373). If this is the case, then 'the hand of YHWH' is also likely a description of the animating force of the spirit. Zimmerli states that different terms which describe the prophet's state of possession are interchangeable.(Zimmerli, W. (1983), *Ezekiel 2*, Philadelphia: Fortress Press, p. 259.

[696] Block, D. I. (1989), 'The Prophet of the Spirit: the Use of RWH in the Book of Ezekiel,' *JETS*, **32** (1), p. 37.

[697] Jacob, E. (1958), *Theology of the Old Testament*, New York: Harper, pp. 161–163.

3. A third level is the transformation of beings and their imbuing with a spirit of rectitude:[698] This is seen in the spirit that drives the Chariot of YHWH, through the Living Creatures for 'wherever the spirit would go, they went' (Ez 1:12); and the wheels: 'for the spirit of the Living Creatures was in the wheels' (Ez 1:21).

4. Fourth is a presence of the Spirit of YHWH: The translocation of Ezekiel through trance and vision or the pouring out of YHWH's Spirit on the Israelites as when Ezekiel is lifted up by the hair (Ez 8:3) or when it is poured out on the house of Israel (Ez 39:29).[699]

Spirit in the Chariot Vision

All four of these levels of spirit appear to have a counterpart within the Chariot of YHWH. These four different manifestations of the spirit: Spirit animating natural phenomena; Spirit as an Animating Force; Spirit of Rectitude and Divine Spirit, will be sequentially followed through their connection with the Chariot.

Spirit as giving Animation to Natural Phenomena

Fire (ēš) and its meaning in Ezekiel

In the Old Testament, fire particularly denotes the divine presence. In particular, the pillar of fire that gave the Hebrews light (Ex 13:21) has analogies with the Chariot of YHWH in its columnar shape, its movement and in its fire. The enlightening and guiding function of the pillar of fire evokes the spirit's function in the Chariot. Fire ēš (אש) is mentioned frequently in Ezekiel -

---

[698] This entails the enfolding of the virtue of prudence which allows the entry of wisdom (see Chapter 3).

[699] This is the only occurrence of the 'pouring on of the divine spirit' which makes *rûaḥ* into a divine fluid that can be poured: Block, D. I. (1998), *The Book of Ezekiel Chapters 25–48*, Grand Rapids: Eerdmans, p.488.

twenty-nine times; this is about the same frequency as in the other major prophets (Isaiah: 34, Jeremiah: 31). A third of these mentions (10) are in connection with the Chariot Vision, so *ēš* is particularly associated with this vision. Fire is not only an earthly sign of combustion but also a spiritual sign of the divine (YHWH appeared to Moses in the burning bush: Ex 3:1-6). Given the complexity of the Chariot Vision, that connects the divine and earthly realms[700] it is important to establish the exact meaning of 'fire' in each occurrence in Ezekiel to hep interpret its occurrence in the Chariot Vision.

The circumstances where fire is mentioned can be divided into several groups: as part of the Chariot Vision, (Ez 1:4 x2; 1:13 x3; 10:6; 10:7); in connection with the 'likeness of a man'[701] (Ez 1:27 x2; 8:2); in anger against Israel (Ez 5:4; 15:4,5,6,7; 19:12,14; 20:47; 21:32; 23:25; 24:10; 28:18; 30:8,14,16; 39:6); as a sacrificial means Ez 20:31 for purification Ez 24:12; to express zeal Ez 36:5; burning coals are mentioned twice: Ez 1:13; 10:2.

In most cases in the book of Ezekiel (16/29) 'fire' is used in the oracles or other chapters to express God's anger against Israel, the consequences of their apostasy, and the threat that they will be consumed.[702] However, in the Scriptures, fire is also used as a method of purification. For example, Nu 31:23 'Everything that can stand the fire, you shall pass through the fire and it shall be clean'. In Ezekiel, it also has other meanings such as a means of sacrifice (Ez 20:31).*[703] The meaning of *ēš* in relation to the Chariot Vision seems to be more

---

[700] Odell, S. (2005), *Ezekiel*, Smyth & Helwys Bible Commentary, Macon: Smyth & Helwys, p. 29.

[701] See section on 'Likeness of a man' image. The 'fire' describing the upper body is a later addition and according to Zimmerli should be removed. (Zimmerli, W. (1979), *Ezekiel 1*, Philadelphia: Fortress Press, p. 88)

[702] 'Like the wood of the vine...which I have given to the *fire* for fuel, so will I give up the inhabitants of Jerusalem. And I will set my face against them: though they escape from the *fire*, the *fire* shall yet consume them.' Ez 15:6–7.

[703] The particular use of the word *ēš* in Ez 36:5 sheds light on the range of meaning of this word in Ezekiel. The word is mostly used in oracles or declarations of anger against Israel. In Ez 36;5, however, YHWH speaks of the fire of his zeal: 'If in *the fire* of my zeal, I have

denoting a presence or source of the spirit. This is because movements of the chariot are subject to the spirit and movements are associated with a lightening-like flash emanating from the fire (Ez 1:14). Also, the fire associated with the 'likeness of a man' is organised anatomically and has to be interpreted on that basis. *Ēš* cannot therefore immediately be ascribed to its use for destruction or judgement, as some commentators see its role (particularly in Ez:10).[704,705]

Fire and lightning in the Chariot Vision

To determine the significance of *ēš* and its potential association with the spirit within the Chariot Vision, I will here review each of the 7 occasions where fire is mentioned in the vision and its association with lightning, (fire associated with the 'likeness of a man' will be dealt with later).

Fire is almost the first thing that Ezekiel sees in the storm cloud (Ez 1:4). In the first mention, fire is described as *wě'ēš mitlaqqaḥat* (וְאֵשׁ מִתְלַקַּחַת) which is literally translated as 'and fire taking hold of itself' but usually translated as 'fire flashing back and forth'. Greenberg says that this denotes a supernatural fire, burning in the air as a fiery mass.[706] Block sees this possibly as lightning bolts.[707]

spoken.' *'im-lô bě'ēš qin'ātî dibbartî* (אִם־לֹא בְאֵשׁ קִנְאָתִי דִבַּרְתִּי): Here YHWH does not speak of anger but of zeal. Although zeal is not anger, elsewhere in scripture, zeal for YHWH is given as a reason for anger, (1 Kg 19:10). It is a burning divine emotion that may precede anger or justified confrontational action. It signifies a deep emotion because the zeal for the honour with which YHWH holds his name is the zeal of the anger that is turned against those who profane it. In Ez 36:22, YHWH goes on to explain that his actions against Israel are a measured response to evil: 'it is not for your sake, O house of Israel, that I am about to act but for the sake of my holy name, which you have profaned'. The action to be taken is to cleanse the house of Israel from idolatry. The meaning of fire in Ezekiel can therefore range between the burning fire of zeal or anger and the fire of purification. *Ēš* appears to have other specific meanings in the Chariot Vision which are explored above.

[704] Taylor, J. B. (1969), *Ezekiel: An Introduction and a Commentary,* London: Tyndale, p. 5.

[705] Zimmerli, W. (1979), *Ezekiel 1,* Philadelphia: Fortress Press, p. 251.

[706] Greenberg, M. (1983), *Ezekiel 1–20,* New York: Doubleday, p. 43.

[707] Block, D. I. (1997), *The Book of Ezekiel Chapters 1–24,* Grand Rapids: Eerdmans, p. 93; Wevers draws a parallel with the pillars of fire and cloud that denoted the presence of

It appears that movement is signified as coming from the fire. Lightning is also described as coming from the fire, *wĕmin haēš yôṣē bārāq* (וּמִן-הָאֵשׁ יוֹצֵא בָרָק) in association with the chariot movements. It was suggested above (Middle Tier Chapter) that 'fire moving itself' (*'ēš mitlaqqaḥat)* as well as the 'lightning' (*bārāq* and *bāzāq*) could be manifestations of the spirit.

There is also a 'brightness surrounding' the cloud from which fire is flashing forth *wenōgah lō' sābîb* (וְנֹגַהּ לוֹ, סָבִיב). The same word, *nōgah* (brightness) being used to describe the fire *wenōgah laēš* (וְנֹגַהּ לָאֵשׁ) later, in Ez 1:13 and Ez 1:27. Both this 'supernatural' brightness surrounding the likeness of a man, and the link to the chariot movement suggest a presence of the spirit.

Spirit as a Live-giving Force

The coals

The first mention of burning coals is in Ez 1:13. Here it says: 'As for the form of the Living Creatures their appearance was like 'burning coals of fire' *kĕgaḥălê ēš bō'ărôt* (כְּגַחֲלֵי-אֵשׁ בֹּעֲרוֹת). Greenberg[708] says that 𝔐 here is corrupt and he reconstructs Ez 1:13 as 'and *amidst the* Creatures was something like burning coals of fire'. The burning coals are a separate entity from the Creatures in Greenberg's interpretation. Burning coals are again mentioned in Ez 10:2 which says: 'go in among the whirling wheels and underneath the cherubim, fill your hands with burning coals from between the cherubim and scatter them over the city'. The coals are the only natural substance of earthly origin in the Chariot Vision. They add a concrete quality to the fire, which otherwise has a spiritual quality with unmeasurable dimensions.[709]

YHWH in Ex 13:21-22. In the wilderness these were symbols of the divine presence for the Hebrews. Cf. Wevers, J. W. (1969), *Ezekiel*, London: Nelson, p. 88.

[708] Greenberg, M. (1983), *Ibid.*, p. 46.

[709] See section below 'Fuzzy Epistemology'. The spirit, fire and coals form a three layered fuzzy epistemological hierarchy that can be analysed using the methodology in the section 'Fuzzy Epistemological Methodology'.

There is a tendency for most commentators to label the burning coals as agents of destruction or of divine justice. For example, Taylor sees in the burning coals (Ez 1:13), that for Ezekiel, fire symbolizing judgement was at the heart of God's presence.[710] Eichrodt says that fire brought from such a holy spot must have a special destructive power.[711] Zimmerli says that the coals scattered over the city are coals of fury as it relates to Ez 9.8 'pour out fury': *bešāpkek 'et- ḥâmâtek* (אֶת-חֲמָתְךָ בְּשָׁפְכְּךָ). [712]

The flaming coals could also be for purification, either of the city once the slaughter had finished or as Cody[713] says, they could be a purification of those who had been marked for sparing. This relates to the burning coal of Is 6:6-7 where the coal, brought from the altar of incense by the seraph, was used to purify Isaiah's lips.[714] Fire is used for cleansing of the 'rust' of lewdness in Ez 24:12.[715] The slaughter by the executioner had finished when coals were scattered and he says 'I have done as you commanded me', referring to the completion of the slaughter (Ez 9:11). This makes it more likely that the flaming coals that are then scattered (Ez 10:2) are for purification of what was remaining. Houk's view is that the coals are for purification and that they are used for marking the foreheads of the faithful.[716] The theme of purification is also suggested by the use of the Hebrew word *zarah* in Ez 10:2 for scattering of the coals: 'and scatter them over the city', *wazerôq al ha'îr* (וּזְרֹק עַל-הָעִיר). In the Old Testament, this word is predominantly used to denote

---

[710] Taylor J. B. (1969), *Ezekiel: An Introduction and a Commentary*, London: Tyndale, p. 56.

[711] Eichrodt, W. (1970), *Ezekiel*, London: SCM Press, p. 134.

[712] Zimmerli, W. (1979), *Ezekiel 1*, Philadelphia: Fortress Press, p. 251.

[713] Cody, A. (1984), *Ezekiel*, Delaware: Michael Glazier, p. 5.

[714] Is 6: 6. 'Behold this has touched your lips; your guilt is taken away, and your sin is forgiven'

[715] Ez 24:12, 'In vain I have wearied myself, its thick rust does not go out of it by fire. Its rust is your filthy lewdness.'

[716] Houk, C.B. (1971), 'The Final Redaction of Ezekiel 10,' *JBL*, **90** (1), pp. 42–54.

sacrificial sprinkling of blood (19/35 instances). For example, Ex 29:20 says 'and sprinkle the blood on the altar': *wazerāqtā et-haddām 'al-hammizbêah sabíb* (וְזָרַקְתָּ אֶת-הַדָּם עַל-הַמִּזְבֵּחַ, סָבִיב).[717]

Another potential link with sacrifice is the use of the word 'flaming torch' *halappidím* (הַלַּפִּדִים) to describe the coals. This is the same word as used in Ge 15.17 to describe the fire that passed between the two portions of Abraham's sacrifice.[718] Ez 1:13 says coals of fire, burning like 'the appearance of torches' *kĕmar'ēh halappidím* (כְּמַרְאֵה הַלַּפִּדִים).[719] Further, on this theme of sacrifice, in Ez 10:6,7, the cherub takes 'fire' that was 'between the cherubim' *'el-haēš ašer bĕynôt hakĕrûbîm* (אֶל-הָאֵשׁ אֲשֶׁר בֵּינוֹת הַכְּרֻבִים). This taking of fire was on the orders of the voice of 'He', presumably God, who commanded the 'man clothed in linen' (Ez 10:6) to 'take fire from between the cherubim'. Interestingly, it is not the man himself who takes fire, but one of the 'cherubim' 'stretches forth his hand' and takes fire and puts it in the hands of this man who takes it and goes out' (Ez 10:7). Also, although the man clothed in linen is told to take 'coals', he is only given 'fire' by the 'cherub'. In the section above on the Living Creatures it is argued that the Living Creatures could be an analogy of the priesthood (see 'Living Creatures significance'). This action of the 'cherub' is in keeping with the Living Creatures having privileged (priestly) access to the coals of fire. An alternative explanation for the coals of fire, therefore, is that they represent

---

[717] The Babylonian Talmud sees a feature of unforeseen mercy in the scattering of coals. 'If the coals from the hands of the cherub had not become cold in the hand of Gabriel (the name given to the cherub who scattered the coals), so would... no remnant and no survivors be left'.

[718] Carley, K.W. (1974), *The Book of the Prophet Ezekiel*, Cambridge: Cambridge University Press, p. 16,

[719] Zimmerli says that this word is a later addition not present in 𝔊. Zimmerli, W. (1979,) *Ezekiel 1*, Philadelphia: Fortress Press, p. 104. 𝔊 in fact gives the same word λαμπάδες (*lampades*) in Gn 15:17, as in Ez 1:13: λαμπάδων (*lampadon*). Neither Allen (Allen, L. C. (1994), *Ezekiel 1–19*, World Biblical Commentary 28, Dallas: Word, p. 7), nor Block (Block, D. I. (1997), *The Book of Ezekiel Chapters 1–24*, Grand Rapids: Eerdmans, p. 95) raise the same concern about this word.

a holocaust offering,[720] around which the priests (Living Creatures) stand. The fire is the blood of the sacrifice that is then taken and sprinkled ('scattered') around. The coals could be an analogy of a sacrificial holocaust and the fire emanating and scattered over Jerusalem could be interpreted as the blood of the sacrifice.[721]

The spirit is central to this interpretation of the coals and of fire. Ez 39:29 says 'And I will not hide my face from them any more when I pour out my Spirit on the house of Israel says the Lord God'. Block says the notion of 'pouring' the Spirit is rooted in the perception of the *rûaḥ* as a sort of divine fluid that covers the object'.[722] This supports that fire represents the pouring of the spirit, as a sacrificial purification by blood. Purification by the spirit is a form of life-giving force.

Spirit of Rectitude that drives the Chariot

The Chariot movements which are described In Ez 1:9 and Ez 1:12, in the description of the Chariot Vision are described as 'each moved straight ahead' (Ez 1:9 and 1:12) or, 'without turning as they went' (Ez 1:9,1:12,1:17). The straightness of these movements were due to the spirit: 'wherever the spirit would go, they

---

[720] Coal, as the dead remnant of a once-living wood, could be an analogy of a dead animal that has been sacrificed.

[721] This sacrifice could be a foreshadowing of the sacrifice of the crucifixion and death of Christ and the pouring out of his blood, which would take place in Jerusalem. Cf. Is 31:9 'the LORD, whose fire is in Zion, and His furnace in Jerusalem.'; Jacob of Serug, Bishop of Serug, Mesopotamia (519), wrote a homily on the Chariot Vision which has been preserved. Jacob sees the Chariot as a portrayal of 'the holy altar of divinity' and the coals signify 'retribution and forgiveness': Golitzin, A. (2003), 'The Image and Glory of God in Jacob of Serug's homily "On that Chariot that Ezekiel the Prophet saw",' *St Vladimir's Theological Quarterly*, **47**, pp. 323–364; p. 355.

[722] Block, D. I. (1998), *The Book of Ezekiel Chapters 25–48*, Grand Rapids: Eerdmans, p. 488: In each of five occurrences in the OT of the pouring out of YHWH's Spirit, a ratification and sealing of the covenantal relationship with his people is signified. Cf: Clines, D. (1968), 'The Image of God in Man,' *TynBul*, **19** (1), p. 82.

went'[723] (Ez 1:12). As discussed above the Living Creatures' faces may reflect psychological forces associated with decision making. The Living Creatures, under the direction of the spirit, could be a metaphor for right reason, and the wheels, which are driven by the spirit of the Creatures, a metaphor for inferior reason. Since superior right reason leads to wisdom, the spirit that directs the Creatures' movements would be a divinely-empowered Spirit.

Divine Spirit in the Chariot

In Ez 8:1 the 'hand of the Lord God' *yad ădōnāy YHWH* (יַד אֲדֹנָי יְהוִה) fell upon Ezekiel and the Spirit of YHWH takes Ezekiel by a lock of hair and he is lifted up by the Spirit. Since the figure described in Ez 8:2, can be identified as the same figure as the 'likeness of a man' that is enthroned on the Chariot in Ez 1:26-27, this passage (Ez 8:1-3) links the 'likeness of a man' in Ez 1:26 with the Spirit of YHWH.

The *ḥašmal* in particular, appears to be a focus of the Spirit. When the 'heavens were opened' (Ez 1:1)[*724] fire was the first manifestation of the Chariot Vision to be seen and it was intimately connected with the *ḥašmal* which was at the centre of the fire (Ez 1:4).[725] In Ez 1:27, fire surrounds the *ḥašmal* seen in the upper part of the body and in Ez 8:2 the *ḥašmal* has 'a brilliant' *zōḥar* (זֹהַר) appearance. This fire and brightness surrounding the *ḥašmal* could specifically, represent the Spirit. As mentioned in 'Upper Tier' *ḥašmal* in the 'likeness of a man' may be identified with the heart of this figure. This highlighting of the

---

[723] Like the English verb 'to go' the Hebrew הלך meaning to go, come or walk, is almost always used intransitively (Strong's Concordance 1980). ללכת therefore does not imply a transitive 'sending' and ילכו would therefore not denote a passive 'transitive' movement, but an action initiated by the Creatures.

[724] Both Origen and Jerome see a parallel between the heavens opening for Chariot and the baptism of Christ.

[725] A link between fire and presence of YHWH is in the burning bush from which YHWH speaks to Moses, Ex 3:2.

heart would fit very well with a central message of Ezekiel YHWH that he will give the Israelites a 'new heart and a new spirit' (Ez 18:31).

Conclusion of the Chapter

The spirit is present in all Chariot Vision levels: In the upper tier, the ḥašmal appears to be a key focus of the spirit as well as the surrounding fire. In the middle tier, the coals, fire and lightning appear to denote the spirit. In the lowest tier the movements of the wheels and of the chariot are spirit-guided.

Chapter 9

Hierarchies and Fuzzy Epistemology

Overview of chapter

This chapter examines known hierarchies in Ezekiel, particularly within the Chariot Vision. A two-level scientific experimental hierarchy and a three-level fuzzy epistemology structure are sought within Ezekiel and applied to Ezekiel and to the Chariot Vision.

Hierarchical thinking is particularly rich and developed in Ezekiel compared to other parts of the Bible.[726] There are two focuses of hierarchy; the Chariot Vision and the Idealised Temple Vision, but hierarchy pervades the whole book of Ezekiel.*[727] In the context of looking for rational structures in Ezekiel it is important to look at hierarchies that are present and their relationship to Ezekiel's message. This chapter will also explore whether a science-based epistemological hierarchy can be demonstrated within Ezekiel.

---

[726] Patton, C. L. and Cook, S. (2004), 'Introduction,' in *Ezekiel's Hierarchical World: Wrestling with a Tiered Reality*, Atlanta: Society of Biblical Literature, pp. 179–198; p. 10.

[727] St Thomas Aquinas sees revelation as a hierarchical operation. 'Supernatural truth reaches us...like the waters of a river which pass over a series of waterfalls from God who is its source to the angels...and then from angels to men.' Gilson, E. (1956), *The Christian Philosophy of St Thomas Aquinas*, Indiana: University of Notre Dame Press, p. 12.

Hierarchies in Ezekiel

The word 'hierarchy' (*ἱεράρχης*) comes from two Greek words: sacred (*ἱερός*) and leadership (*ἀρχής*).[728] The structure in hierarchical thinking therefore originates in sacred levels of authority with God at its highest rank. Hierarchies are therefore natural structures in the Bible.[729] The most prominent in Ezekiel, is the sacred space hierarchy that is contained within the Idealised Temple Vision. Block shows that there are 11 different levels of sacrality as one moves horizontally from east to west in the reconstructed temple plan.[730] The sanctuary space has gradations of holiness as part of the sanctuary system.[731]

In the Chariot Vision there are several overlapping hierarchies. Historically the vision has been viewed as a tiered structure with three main constituents: the upper one with the likeness of a man surrounded by fire; below this, separated by a firmament of crystal are four Living Creatures; the lowermost tier is the wheels.[732] Cook points out that Ezekiel's language gives a vertical organisation, by use of the words 'above' *'al* (על) in Ez 10:1, to describe where the throne and likeness of man were and 'below' *tahat* (תחת) to describe where the wheels are, in relation to the Living Creatures.[733] The Vision contains other hierarchies:

---

[728] Hierarchies are usually rational in their structure. Although hierarchy can also refer to an informal structure of inequality in power, such as class structure in society and hegemony in world politics, these are probably distortions of their original sacrality-based structure. Cf. Miura, S. (2007), 'Hierarchy,' *Encyclopedia Britannica*, at https://www.britannica.com/topic/hierarchy-social-science.

[729] Olyan, S. M. (2000), *Rites and Rank: Hierarchy in biblical representations of cult*, Princeton: Princeton University Press, p. 3.

[730] Block, D. I. (1998) *The Book of Ezekiel Chapters 25–48*, Grand Rapids: Eerdmans, p. 571.

[731] Olyan, S. M. (2000), *Ibid.*, p. 22.

[732] Joyce, P. M. (2007), *Ezekiel: A Commentary*, New York: T & T Clark, p. 68.

[733] Cook, S. L. (2004), 'Cosmos, *Kabod* and Cherub: Ontological and Epistemological Hierarchy in Ezekiel,' in *Ezekiel's Hierarchical World: Wrestling with a Tiered Reality*, C. L. Patton and S. L. Cook (eds), Atlanta: Society of Biblical Literature, pp. 179–198; p. 185.

1. A cosmic hierarchy with the transcendent spirit at the top, the Living Creatures that have human bodies, but also angel-like wings at a second level, and the wheels, a human construct and in contact with the earth, delineating the lower limit of the Chariot's cosmic extent.

2. A hierarchy of degree of holiness with the Glory of the God of Israel seated on the top, the Living Creatures who are suffused by the spirit in the centre,[734] and the wheels which are guided by the spirit of the Creatures below. The wheels still show a link with the divine through their eyes but are less directly in touch with the spirit and contact the unholy Babylonian ground.

3. A horizontal ontological hierarchy:[735] at the centre of the Chariot is the numinous energy of the transcendent fire of the spirit which is between the wheels of the Chariot (Ez 1:13). Surrounding this, are the Creatures who have both human and divine characteristics, and thirdly God's people as a horizontally orientated hierarchy. The wheels are junctional structures, that link God and his people through their movements.

4. An epistemological hierarchy which Cook sees as a hierarchy of the human knowledge of God.[736] At the upper end, Ezekiel experienced God's direct communication 'the spirit lifted me up' (Ez 3:12) an indication of direct special revelation that Ezekiel transmitted to the Hebrew people. At the lower end, the archetypal form of the Creatures[737] gives them symbolic power to communicate theological truth to foreign peoples who have no direct access to Israel's covenant. This is a two- level hierarchy in which the wheels are not mentioned.

---

[734] Launderville D. F. (2007), *Spirit and Reason*, Waco: Baylor University Press, p. 374.

[735] Cook, S. L. (2004), *Ibid.*, p.182.

[736] Cook, S. L. (2004), *Ibid.*, p. 196.

[737] Rahner, K. (1978) *Foundations of Christian Faith: An introduction to the Idea of Christianity*, (trans. W. Dych), New York: Crossroad, p. 81.

5. A degree of wisdom hierarchy can be discerned in the Chariot Vision with *sapientia* (wisdom) at the top in the figure of the Glory of the God of Israel; *ratio superior* (superior reason) represented by the Living Creatures: *ratio inferior* (inferior reason) and *scientia* (knowledge or science) represented by the wheels as discussed in this thesis.

6. A hierarchy of symbolism can also be found across Ezekiel, going progressively from metaphor, to metaphoric narratives, sign-actions (more properly called symbolic acts), and the most detailed symbolism in his visions. The Chariot Vision and the Idealised Temple Vision contain the most extensive and detailed symbolism. (see 'Metaphor and Symbolic Reasoning')

Science-based Hierarchies

Scientific Reasoning and Ezekiel

Since science is a form of reasoning, it would fit in the Book of Ezekiel, as a natural extension of Ezekiel's propensity to reason and measure. As discussed above, science is a form of inferior reason as its goal is to understand the spatio-temporal world by using reason. Gaining knowledge in science is 'exploratory' as opposed to 'revelatory' in Theology.[738] Scientific method includes experimentation, observation (or measurement) and theory construction.[739] Science is *deductive* or *inductive* depending on whether the conclusions are contained in the known facts or not. There are four steps in a scientific experiment:[740]

1. Formulation of a hypothesis or theory.

---

[738] Rheinberger, H.-J. (2005), 'A Reply to David Bloor: "Toward a Sociology of Epistemic Things",' *Perspectives on Science*, **13** (3), pp. 406–410; p. 409.

[739] Okasha, S. (2016), *Philosophy of Science*, Oxford: Oxford University Press, p. 2.

[740] Cf: Hausman, D. (1992), *The Inexact and Separate Science of Economics*, Cambridge: Cambridge University Press, p. 304.

2. Deduction of a prediction or of an observable or measurable aim from the hypothesis, which tests the hypothesis.

3. Testing the hypothesis using an experimental system.

4. Judging whether the hypothesis is confirmed or disconfirmed based on whether the prediction turns out to be true or false.

A particular form of inductive reasoning called Inference to the Best Explanation (IBE) is used in day-to-day mental reasoning and in science. It starts with an incomplete or imprecise observation, and reasons from this to a theory or hypothesis that explains the data.*[741] In Ezekiel, for example, IBE would be used in arriving at a conclusion about the meaning of the reconstructed temple: precise measurements are available and give us a picture of the temple; concluding that it is 'a rational theocentric exposition of Israel's future' [742] is the best explanation based on these measurements but cannot be proven.

Hierarchy of a Scientific Experiment

A scientific experimental system is used to test a hypothesis and determine whether it is true or false. It almost always involves measurement of some kind[743] and the production of mathematical data.[744] In structuralist terms, an experimental system has a two-level hierarchy: the theoretical and the empirical (Figure 18). The theoretical level is the hypothesis that must be proven or disproven. For example, faced with the vision of the water flowing from the Temple (Ez 47:1), the hypothesis of the scientist (Ezekiel) might be: 'this is a

---

[741] A good example of IBE is Darwin's theory of evolution which was put forward as the best explanation for many species of animals with close anatomical similarities.

[742] Block, D. I. (1998), *The Book of Ezekiel Chapters 25–48*, Grand Rapids:, Eerdmans, p. 746.

[743] Popper, K. (1972), *The Logic of Scientific Discovery*, London: Hutchinson, p. 126.

[744] Sismondo S. (2009) *An Introduction to Science and Technology Studies*, Chichester: Wiley, p.85.

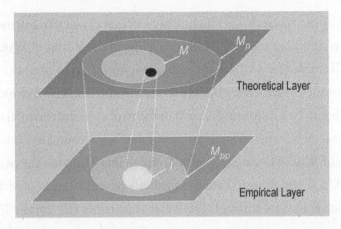

Figure 18: Structuralist view of an experimental system with theoretical and empirical layers. Source: Seising, R. (2010), 'Fuzzy Sets and Systems and Philosophy of Science,' in Seising R. (ed.), *Views on Fuzzy Sets and Systems from Different Perspectives*, Hamburg: Springer, p. 18, (Fig 1.8).

river'. To prove or disprove the hypothesis, the depth of the water is measured at four specific intervals (this is the empirical level). In view of the exponential increase in the amount of water that Ezekiel registers, the hypothesis is disproven: the conclusion is that this is not an ordinary river. The theoretical level is therefore the object of inquiry that the experiment is designed to prove or disprove; the empirical level is the actual experimental process.

In philosophical terms, these two levels of the experimental system are called the epistemic and the technical. The term 'epistemic object' was coined by Rheinberger to denote the researcher's abstract object of inquiry that is the driving force of the experimental system.[745] It denotes the hypothesis, that is

---

[745] Rheinberger, H.-J. (1991), 'The "epistemic thing" and its technical conditions – from biochemistry to molecular biology,' in Gremmen, B. (ed.), *The Interaction between Technology and Science*, Netherlands: Wageningen (NL): Wageningen Agricultural University, pp. 281–298.

the target of the research. [746] Technical objects are concrete instruments,[747] such as measurement tools: the measuring line in the Vision of Water flowing from the Temple (Ez 47:3) or the measuring reed (Ez 40:3) in the Idealised Temple Vision. The concrete objects that are measured are also technical objects, and there has to be a measurable object in the two-level experimental system.[748]

The two-level hierarchy of an experimental system is found in Ezekiel, within the Vision of Water Flowing from the Temple. This vision has the two essential elements:[749] firstly, the unstated hypothesis (the epistemic object) or why the man leads Ezekiel through the water four times (Ez 47:3). This might be: 'the flowing water is a normal river'. Secondly, the technical objects of measurement: (a) the line that measures distances of 1000 cubits, (b) Ezekiel's body, by which the depth of the water is measured and c) the river which is measured.[750]

The abstract/concrete relationship that is reflected in this epistemic/technical dyad is key in science. The presence of this science-related structure in the Vision of Water Flowing from the Temple suggests that science is not alien or contrary to the Bible and can be a tool of divine revelation. The Vision of Water Flowing from the Temple illustrates that while science can give rational conclusions about the world around us (the river) it is unable to measure the 'transcendent' (exponential increase in the river). The two-level scientific structure in the Vision of Water Flowing from the Temple is likely uncommon in

---

[746] Ewenstein B., Whyte, J. (2009), 'Knowledge Practices in Design: The Role of Visual Representations as "Epistemic Objects",' *Organisation Studies,* **30** (1), pp. 7–28; p. 9.

[747] See Appendix 3: A comparison of epistemic, boundary and technical objects.

[748] The realist approach to science sees the experimental system as discovery, but it may also be interpreted as invention or construction. Rheinberger, H.-J. (2005), 'A Reply to David Bloor: "Toward a Sociology of Epistemic Things',' *Perspectives on Science,* **13** (3), pp. 406–410; p. 410.

[749] Wartofsky, M. W. (1968), *Conceptual Foundations of Scientific Thought,* New York: Collier-Macmillan, p. 181; 'Experiment' means experimental methodology, particularly through measurement, to reach a conclusion by inductive reasoning.

[750] The river is also part of the empirical layer as it is measurable near its source.

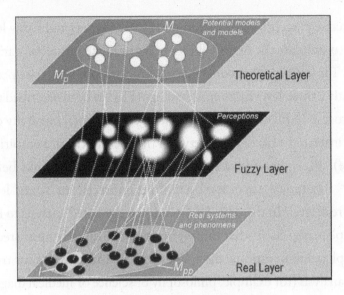

Figure 19: Seising's Fuzzy Epistemology.

the Bible, however, Cook shows that other two-level epistemological structures exist in Ezekiel,[751] and the two-level experimental system hierarchy fits in with other binary hierarchies in the Bible.[752] I will now look at a three-level modification of this two-level experimental system structure and discuss whether it can be found in Ezekiel or the Bible.

A three-level science-based epistemological hierarchy

Symbolism is used frequently in Ezekiel and Launderville has shown that symbolic reasoning plays a central role.[753] The brain preferentially uses symbols

---

[751] Cook, S. L. (2004), 'Cosmos, *Kabod* and Cherub: Ontological and Epistemological Hierarchy in Ezekiel,' in *Ezekiel's Hierarchical World: Wrestling with a Tiered Reality,* C. L. Patton and S. L. Cook (eds), Atlanta: Society of Biblical Literature, pp. 179–198; p. 196.

[752] Olyan, S. M. (2000), *Rites and rank: Hierarchy in biblical representations of cult,* Princeton: Princeton University Press, p. 13.

[753] Launderville, D. F. (2007), *Spirit and Reason,* Waco: Baylor University Press, p. 349.

and imprecise perceptions to reason with.[754] Fuzzy sets and fuzzy logic were invented to simulate the brain's ability to manipulate imprecise data and perceptions. The brain reaches concrete conclusions from imprecise perceptions and recently a three-level fuzzy epistemology (FE) has been described that mimics this process.[755] FE modifies the two-level experimental hierarchy described above by inserting a fuzzy layer between the theoretical and empirical layers (Figure 19). FE is therefore a science-based structure that links between an abstract[756] theoretical upper level, a middle, imprecise, or 'fuzzy' level and a concrete 'real' level. In this way, FE extends science-based analysis to imprecise or incomplete objects,[757] that cannot normally be directly measured. FE has found application in rational disciplines that are too complex for traditional scientific analysis (for example, philosophy of science or medical diagnosis).[758]

FE is similar to the process of symbolic reasoning[759] as both use amodal (incomplete or imprecise) markers to represent symbols or perceptions. As symbols participate in the reality to which they point, a fuzzy object contains

---

[754] Landy, D., Allen, C., Zednik, C. (2014), 'A perceptual account of symbolic reasoning,' *Frontiers in Psychology*, 5, p. 1.

[755] Seising, R. (2010), 'Fuzzy Sets and Systems and Philosophy of Science,' in Seising, R. (ed.), *Views on Fuzzy Sets and Systems from Different Perspectives*, Hamburg: Springer, p. 25.

[756] Entities that cannot be perceived or measured (such as an abstract idea) or a spiritual entity.

[757] Such as natural phenomena (fire, lightning). Mayr identified several characteristics of Biology that lend themselves to a fuzzy analysis rather than a traditional scientific analysis: (a) unsharp separation of classes of phenomena, (b) the occurrence of variation or chance events, (c) systems that are highly complex and d) the absence of strict underlying rules. Mayr, E. (2004), *What Makes Biology Unique? Considerations on the autonomy of a scientific discipline*, Cambridge: Cambridge University Press, p. 4.

[758] De Carvalho, L. M. F., *et al.* (2008), 'A neuro-fuzzy system to support in the diagnostic of epileptic events and non-epileptic events using different fuzzy arithmetical operations,' *Arq Neuro-Psiquiatr*, 66 (2A), pp. 179–183.

[759] Landy, D., Allen, C., Zednik, C. (2014), *Ibid.*, 'A perceptual account of symbolic reasoning,' *Frontiers in Psychology*, 5, p. 1. for example, 'Semantic Markerese' is a postulated intermediate brain language like GCL.

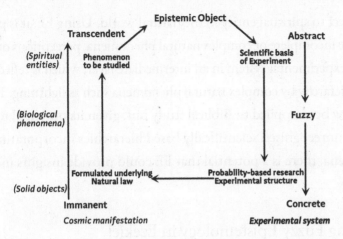

Figure 20: Cosmic and Fuzzy Epistemology Hierarchies Compared.

within it, the concrete object to which it is linked in 'fuzzy form' – (in FE, there must be an epistemological link between the object of the fuzzy layer and the object in the real layer: see Figure 19). FE is therefore a science-based structure that can incorporate 'symbols' of complex or incomplete phenomena. Given the frequent use of symbolism in Ezekiel an attempt was made to determine the presence of any 3-level FE structures in Ezekiel.

Figure 20 shows a comparison of the three levels of fuzzy epistemology on the right, compared with a cosmic hierarchy on the left. The inner circular path traces the progress of a classical scientific experiment. It starts with the epistemic object, which is the hypothesis or aim of the experiment. Next the structure of the experiment (the scientific basis of how the hypothesis will be tested) is decided. Thirdly, an experimental structure is set up and the chosen object to be investigated is measured using a concrete measuring apparatus. Fourthly, a conclusion is reached and if the hypothesis is confirmed it may lead to formulation of an underlying natural law. Fifth, this law can help determine a further phenomenon to be studied. The progression from abstract to concrete in the experimental system is mirrored by a cosmic hierarchy of solid object to spiritual entity in the natural world. Using FE, it is possible to introduce incomplete or complex natural phenomena, perceptions or language into the experimental system in an intermediate layer, which is reflected in the cosmic hierarchy by complex natural phenomena such as lightning. FE has not up to now been applied to Biblical study but, given its potential to discover hitherto unrecognised scientifically based hierarchies incorporating natural phenomena, there is a potential that FE could provide insights in the book Ezekiel.

Applying Fuzzy Epistemology in Ezekiel

Many of the theophanies in Scripture (see Appendix 4) are manifested by natural phenomena such as fire, thunder, lightning flashes and clouds,*[760] whose appearance could be analysed in a similar way to characteristics cited above: highly complex, chance events which are not separated sharply by class and do not appear to obey strict rules.[761] The Chariot of YHWH vision of Ezekiel has several complex natural elements: flashing fire (Ez 1:4), lightning (Ez 1:14), a rainbow (Ez 1:28), sounds of waters, thunder (Ez 1:24), or of an earthquake (Ez 3:12), burning coals (Ez 10:2). This suggests that FE may have a place in the analysis of the Chariot of YHWH vision and that the three-level FE structure might be found within the Book of Ezekiel.

Detection of Fuzzy Epistemology structures: methodology

Essential elements for the presence of an FE structure are: (a) The three levels of FE must be present and must be directly connected epistemologically. (b) The top level should be abstract or spiritual; (c) the middle level must be a

---

[760] Of 18 theophanies in Appendix 4 marked with an asterisk, the following are the main 'biological' manifestations: cloud 11; fire 9; earthquake 3; smoke 3; thunder 2; lightning 2; hail 1; rainbow 1.

[761] Mayr, E. (2004), *What Makes Biology Unique? Considerations on the autonomy of a scientific discipline,* Cambridge: Cambridge University Press, p. 4.

perception, language or complex natural phenomenon; (d) the lower level must be a concrete object that is objectively measurable. This methodology will be applied to a sample text from Ezekiel and a further Biblical text, to determine (a) if the methodology is practicable, (b) if an FE structure can be found in this text and (c) if the exercise reveals any new insights about the text.

A Bible passage is selected. It must contain 3 levels:

1. Upper level: Abstract or spiritual concept that cannot be perceived or measured. (Epistemic Object)

2. Middle level: Perceptions or conceptions of objects that are fuzzy, like natural or biological phenomena, language or symbols that relate to a concrete object.

3. Lower level: A concrete object. Must be discrete, directly measurable objects (or measuring tools, e.g.: scales, measuring rod). A fixed, counted number of objects is also acceptable.

The passage is reviewed and all entities and objects are classified into one of the 3 levels. If objects are combinations of these levels or do not easily fall into one of these three, they should not be included. One item from each level is selected to form an FE structure. The items selected should have a connection with each other using a hypothesis that links the upper and lower levels. The hypothesis should be rationally based and consider the real object, its measurement or its empirically definable qualities.

The science or rational-based role of the selected concrete object in the context of the established FE structure is looked for. What does it tell us about the uppermost level: the epistemic object? Does it give us any rational-based insights? An attempt to reach a conclusion is made.

If the hypothesis is broadly confirmed and there are new rational based insights, this would tend to confirm the validity of the selected FE structure.

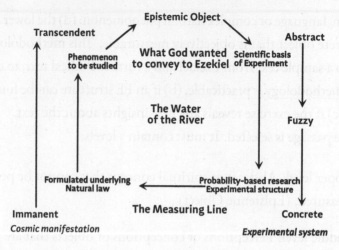

Figure 21: Fuzzy Epistemology Structure of Vision of Water flowing from the Temple.

Application to the Vision of Water flowing from the Temple

The above methodology will now be applied to a Biblical text. The passage has been chosen from Ezekiel because of its contextual relevance to the Chariot Vision. A vision was chosen since visions often are difficult to interpret and would potentially benefit from a new method of interpretation. The Vision of Water flowing from the Temple (Ez 47: See Appendix 5 for text) was chosen because of the combination of a natural phenomenon (the river) and measurement, which appeared compatible with the three layers of FE experimental system. The questions to be asked are: (a) Is it feasible to use this methodology to analyse a Biblical text such as this? (b) Can the three levels of the methodology be discerned satisfactorily? (c) are there any insights or conclusions to be reached? From this Ezekiel text (Ez 47:1-12) three levels selected are (see Appendix 7a and Figure 21):

Results

1. Upper level: What God wanted to convey to Ezekiel (the epistemic object). For example, that this was not a normal river.

2. Middle level: the exponentially expanding waters of the river.

3. Lower level: the measuring line used by the man. (Note that Ezekiel's body is used as a separate measuring tool for the depth of the water. This is not chosen for the current analysis but could be the lower level of a further analysis).

A hypothesis for this 'experiment' might be that objective measurement (since a measuring line is the 'Concrete Object') can allow insights into God's will for us.

The conclusion in this case could be that objective measurement (or 'science') allows us to understand up to a certain point (Ezekiel is able to cross the river at 1000, 2000, and 3000 cubits distance but not at 4000 cubits), but there is a limit to the ability of the human mind to comprehend even with the help of science (inability to cross the river at 4000 cubits distance).[762]

Application to Revelation 6.5: The Black Horse of the Apocalypse

The vision of the Black Horse of the Apocalypse (Re 6:5–6, Appendix 6 for text) was chosen because it has elements of measurement in it and its meaning is not clear. As for the Ezekiel water vision, the same three questions are asked: (a) Is it feasible to use this methodology to analyse a biblical text such as this? (b) Can the three levels of the methodology be discerned satisfactorily? (c) are there any insights or conclusions to be reached?

---

[762] This Vision has similarities with the famous experiment of Archimedes See 'Lower Tier': Comparison of Ezekiel's (623 BC) and Archimedes's (288 BC) experiments.

Results

Three levels were selected from the available objects (See Appendix 7b and Figure 22):

1. Upper level: The intended meaning of the Black Horse Vision.

2. Middle level: For this level is chosen *what* the voice says: '*A quart of wheat for a day's pay, and three quarts of barley for a day's pay but do not damage the olive oil and the wine!*' Spoken words are an accepted form of fuzzy knowledge.[763]

3. Lower level: The pair of scales. Scales are a measuring tool for weighing.

The hypothesis might be that since the scales are a measuring tool, the reason for the black horse is related to measurement or science. Looking at the middle level, accurate weighing (in quart measures) appears to be important for the wheat and the barley but not for the oil and wine (which are not measured). In the context of the Book of Revelation, the scales (which have the significance of application of justice) may also point to judgement at the end of time.

A conclusion based on this could be that we must ensure an appropriate payment for work by using objective measurement. A second conclusion could be that we will be judged on our fair dealings with employees. The lack of need to measure the oil and wine may be because these represent kindness to employees that can't be measured. A more general conclusion could be that science and justice should work together but should not exclude mercy.

In summary therefore it is feasible to apply the FE methodology to these passages, and it did appear to produce valid insights. FE does appear to be a practicable form of scriptural analysis in specific situations. These passages were not randomly selected however, and further testing would have to be

---

[763] Zadeh, L. A. (1996), 'Fuzzy Logic = Computing with Words,' *IEEE Transactions on Fuzzy Systems*, 4 (2), pp. 103–111; p. 103.

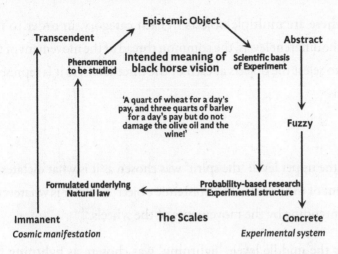

Figure 22: Fuzzy Epistemology Structure of Black Horse Vision.

carried out to see where this method is applicable. It is likely to be usable in relation to theophanies, such as those listed in Appendix 4, although the Vision of Water Flowing from the Temple text used is not a theophany. The methodology therefore does appear to be applicable to the Chariot of YHWH Vision.

Applying Fuzzy Epistemology to the Chariot of YHWH Vision

Fuzzy epistemology in the Chariot Vision

The methodology set up in the previous chapter will now be applied to the Chariot of YHWH Vision. Appendix 8 gives the entities and objects that are mentioned in association with the first detailed appearance of the Chariot of YHWH in Ez 1:4. They are classified into 3 groups according to the methodology above. Those that do not fit into any of the three groups are put under 'unclassifiable', which is that they do not relate clearly to any individual FE level.

Since there are multiple objects in each category, in order to find a link between the different levels, the common thread of the movement of the chariot was used to select the objects at each of the three levels, as it is represented in all three levels.

Results

1. In the upper level: 'the spirit' was chosen as it is what dictates the movement of the wheels. The 'epistemic object' however is whatever the spirit is intending by the movements of the wheels.[764]

2. For the middle level: 'lightning' was chosen, as lightning flashes are linked in Ezekiel to the movement of the wheels,[765] see Figure 23.

3. For the lower level: the choice for a concrete object is 'wheels' as they are linked to the movement of the Chariot, and the spirit dictates their movements. Although we are dealing with visionary objects, unlike the other objects in the Chariot Vision, wheels are structures that are normally material and measurable.

The three objects chosen, appear to be congruent with a scientific experimental system epistemic object (upper level), a fuzzy object (middle level) and a concrete object (wheels: lower level).

The fact that it is possible to arrange objects that relate to the movement of the Chariot of YHWH in an FE experimental system hierarchy suggests: (a) that there is an inherent epistemological hierarchy of the objects in the Chariot connected with its movement;*[766] (b) that the Chariot is exhibiting a

---

[764] Ez 1:20 'Wherever the spirit would go, they (the wheels) went'.

[765] Ez 1:14 'Out of the fire went forth lightning. And the Living Creatures darted back and forth like a flash of lightning...and when the Living Creatures went the wheels went with beside them'

[766] Since 'coals' are also concrete and measurable, a similar three-level fuzzy epistemological structure can be obtained using: upper level: what the 'man' wished to achieve by

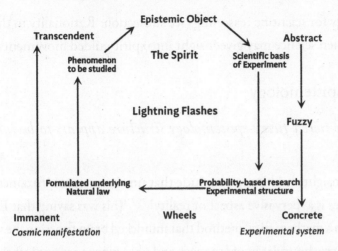

Figure 23: Fuzzy Epistemology Structure of Movements of the Chariot of YHWH.

progressive rational-based revelation concerning its movements by means of its organisation. The fact that a wheel is a metrological tool and that measurement is central to science, may be relevant to its role in the context of this FE structure. Although we are not given any measurements of the distances covered by the movements of the Chariot, the wheels suggest that distance is an important part of the rational-based revelation. Also since the wheel moves in different directions and changes speed, issues of space and of time are also strongly implied. What these movements could mean is discussed in the following chapter.

The conclusions therefore are that it is feasible to analyse the Chariot of YHWH Vision movements by FE. Secondly that the organisation of the movements of the Chariot, are congruent with a layered epistemological hierarchy that mimics a scientific experimental system. Thirdly the three levels of the Chariot show a progressive rational-based revelation concerning its movements. In terms of a specific conclusion for this FE structure, if the wheels are taken as

---

ordering scattering of coals (Ez 10:6); middle level: fire between the wheels; lower level: 'coals'.

an analogy for scientific reasoning (see subsection: Rationality in the Chariot Vision) then science may give insight into spirit-guided movements.

Fuzzy Epistemology

The three-tiered fuzzy epistemology structure appears to be inherent in nature

Zadeh's insights led him to conclude that un-sharpness or 'fuzziness' of class boundaries is a 'pervasive aspect of reality'.[767] This was saying that FE is much more than a mathematical method that mimicked brain function, he was saying that the fuzzy boundaries of language and phenomena are not just the way our brain perceives a precise natural world: imprecise boundaries are an objective part of the world around us. In the foregoing section it was shown that biology, unlike traditional sciences, has highly complex, chance events which are not separated sharply by class and do not appear to obey strict rules.[768] We are all too accustomed to thinking of 'reality' as the world of structure and technology that has been built up around us. However, Zadeh is telling us that the real world, unlike the technological world, has unsharp or fuzzy boundaries.[769]

Potential links between Babylonian Science and Fuzzy Epistemology

The fact that the Babylonians developed mathematical predictive models for the positions of the stars and planets and used an internal logic in arriving at their results shows that they were the first to start to use what is now accepted as scientific methodology. Whether what they did represents science is debatable,

---

[767] Singh, H. *et al.* (2013), 'Real-Life Applications of Fuzzy Logic,' *Advances in Fuzzy Systems,* at https://doi.org/10.1155/2013/581879.

[768] Mayr, E. (2004), *What Makes Biology Unique? Considerations on the autonomy of a scientific discipline,* Cambridge: Cambridge University Press, p. 4.

[769] Heller, quoting Leibniz, says that God thinks in a way that is analogical to humans' mathematical thought: Heller, M. (2008), 'Deciphering the Mind of God,' *First Things,* at *https://www.firstthings.com/web-exclusives/2008/03/deciphering-the-mind-of-god.*

however their computations did lay the groundwork for the development of early science.

The fact that the Babylonians (a) thought that the movements of the planets were understandable and therefore not random, and the fact that (b) they tried to develop ways of predicting these apparently random movements shows that they subscribed to a rational view of the world, which is essential for the development of science.

Fuzzy mathematics replaces uncertain information by using sets and systems and in this way allows the uncertain information to be processed. This in essence, is what the Babylonian astronomers were doing to the uncertain movements of the planets.

In conclusion, this chapter has reviewed known hierarchies in Ezekiel and looked for a science-based hierarchy (FE) within three passages. Within the Chariot Vision, FE is able to uncover a progressive rational based revelation concerning its movements. Since rational organisation appears to be inherent in the Book of Ezekiel, this science-based structuring does fit with the Ezekiel's rational way of thinking. The implications of this for Ezekiel and for science are discussed further in Chapter 12.

Conclusion of the Chapter

A two level scientific hierarchy is detectable within the Vision of the Water Flowing from the Temple. Three-level fuzzy epistemology hierarchies were found in a Bible passage and in the Vision of the Water Flowing from the Temple passage. The finding of a fuzzy epistemology hierarchy within the Chariot Vision structure suggests that it exhibits a progressive science-based revelation.

Chapter 10

Movements of the Chariot

Overview of the chapter

This chapter examines the vertical and horizontal movements of the Chariot and their potential significance. Two principal interpretations are given for the key movements made by the Chariot: the Idealised Temple and the Christo-centric interpretations. The rapid, lightning-like movements are also discussed.

Vertical Movements and Astral and Solar orientation

Since the Living Creatures are a link between the divine and the earthly, it suggests that the four character-forces represented by the Creatures may be used by God in his communication with man. Maimonides says that the account of the Chariot is identical with divine science.[770] In saying this he appears to be saying that the Chariot Vision is a structure that reveals *divine* information in a rational way.[771] Two findings from Ezekiel appear to support this. On

---

[770] Moses Maimonides (1963), *The Guide of the Perplexed*, Chicago: University of Chicago Press, p. 6.

[771] Astronomy is part of Divine Science and the Creatures in particular may have this function as Maimonides links them to celestial spheres.' Kriesel, H. (2015) *Judaism as Philosophy: Studies in Maimonides and the Medieval Jewish Philosophers of Provence*, Boston: Academic Studies Press, p. 248.

the basis of the Hebrew calendar information given by Ezekiel, it is possible to calculate the exact date of the initial appearance of the Chariot of YHWH (Ez 1:1) to 31 July 593 BC[772] and that of its final appearance in the Idealised Temple Vision (Ez 40:1) to 28 April 573 BC.[773] The initial vision therefore took place in mid-summer, the season expressed by the face of the Lion, the time of the year when Babylonian kings would go out to war. The appearance of the Chariot at this time with its arresting vision of the Glory of the Lord (Ez 1:28) is totally appropriate to the 'authoritativeness' attributed to the Lion face and suggests that this was not a mere coincidence. Likewise, the final vision took place in spring, the season expressed by the Bull/Cattle as the time of year of beginnings and hope. Here YHWH makes his divine speech (Ez 43:6-12) which proclaims the holiness and mercy of God despite the sinfulness of humans and is an important element in understanding the Chariot Vision.[774] The celestial timing of this appearance is again totally appropriate to the 'nurturance' expressed by the Bull/Cattle. The final vision took place nearly 20 years after the first one. This long passage of time could suggest that time will pass before the fulfilment of what the Idealised Temple Vision contains, and therefore could be a sign of delayed fulfilment. This is also suggested by numerology of the number of days of months that elapse (See 'Numerology')

These two occurrences support that the different faces of the Creatures may reflect different moods of YHWH by the season in which he chose to act. They also give an illustration of the rational and hierarchical nature of the Chariot structure, particularly the Living Creatures and its connection with the cosmos. That character-forces of YHWH displayed through nature could be understood by man would illustrate how man is made in the likeness of God (Ge 1:27).

---

[772] Block, D. I. (1997), *The Book of Ezekiel Chapters 1-24*, Grand Rapids: Eerdmans. p. 83.

[773] Block, D. I. (1998), *The Book of Ezekiel Chapters 25–48*, Grand Rapids: Eerdmans, p. 513.

[774] Block, D. I. (1998), *Ibid.*, pp. 580, 590.

In terms of solar orientation, the initial Chariot Vision (Ez 1:4) came from the north.[775] For the Jews, the north was the direction of confrontation (Ez 26:7). YHWH was coming to deliver a stern message through Ezekiel, whose tone is often intentionally confrontational because of his hardened audience (Ez 3:9); it was fitting therefore, that the inaugural vision came from the north. The second arrival of the Chariot was at the north gate of the Temple (Ez 8:4). This was just preceding the Visions of the Abominations in the Temple and that of the Executioners. Here again the direction from the north was very appropriate.

Once the execution had taken place and fire had been scattered over the city, the Glory of God left by the East Gate, having departed from the Temple, and stood on the mountain on the east side of the city. The departure to the east is a conveying of the Glory to its real abode in the heavens.[776] The east is the direction from which the Chariot reappears twenty years later, when it enters the temple (Ez 43:2) and in his mercy YHWH re-established his throne (Ez 43:12).

The solar or compass orientation of the vertical movements of the chariot tell a story in addition to that of the astral orientation. These vertically directed movements synchronised in time, season, direction and compass orientation, coming from the clouds and leaving to the hills, speak of the transcendent sovereign power of YHWH, as a divine king enthroned on his chariot and are a metaphor for divine rule.[777]

---

[775] This is unusual as in Mesopotamia storms come from the southwest – Brownlee, W. H. (1986), *Ezekiel 1–19*, World Biblical Commentary 28, Waco: Word, p. 10.

[776] Block, D. I. (1997), *The Book of Ezekiel Chapters 1–24*, Grand Rapids: Eerdmans, p. 358.

[777] Launderville, D. (2004), 'Ezekiel's Throne-Chariot Vision: Spiritualizing the Model of Divine Royal Rule,' *Catholic Biblical Quarterly*, 66 (3), pp. 361–377; p. 376.

Horizontal Darting Movements and Decision Making

Although the Chariot's vertical movements incorporate horizontal progression, the Creatures also make lightning-like, to-and-fro, purely horizontal movements. The Chariot has wheels but Eichrodt says the wheels are only a symbol for motion, since movement of the Chariot is provided by the 'might of the Spirit'.[778] Cook says that 'the cherubim permit God's saving power to project out onto the land of Israel, creating a holy territory for God's people.'[779]

These darting movements are accompanied by lightning-like flashes from the fire between the Creatures (Ez 1:14). The fire and its associated lightning therefore are intimately associated with the Creatures' movements, which in turn are associated with movement of the wheels and the whole chariot structure. Ez 1:20 says that the Living Creatures had the spirit within them and this spirit was also in the wheels; the wheels would only go where the Living Creatures went.

Since the creatures have will, their intellects have to be perfectly coordinated to perform these straight, lightning movements. As discussed above under 'Moral Decision Making' these movements are therefore as much a metaphor for intellectual activity as they are for real movement,[780] and can equate to a 'controlled mental contribution' in human decision making.[781]

---

778 Eichrodt, W. (1970), *Ezekiel*, London: SCM Press, p. 57.

779 Cook, S. L. (2004), 'Cosmos *Kabod*, and Cherub: Ontological and Epistemological Hierarchy in Ezekiel,' in, *Ezekiel's Hierarchical World: Wrestling with a Tiered Reality*, C. L. Patton and S. L. Cook (eds), Atlanta: Society of Biblical Literature, pp. 179–198; p. 187.

780 Lefebvre (Lefebvre, H. (1991), *The Production of Space*, Oxford: Blackwell, p. 299.) sees that space can be 'representational'; Cf. also Knott, K. (2005), *The Location of religion: A Spatial Analysis*, London: Routledge, p. 128. for the metaphorical use of space; Cf also Smith, J. Z. (1987), *To Take Place: Towards Theory in Ritual*, Chicago: University of Chicago Press, p. 117. Smith shows how Ezekiel uses external space as a metaphor for human internal mental space.

781 As the different faces of the Living Creatures are taken as metaphors for different psychological traits, also the Chariot's rapid movements are taken as a metaphor for decision

Significance of the Movements of the Chariot of YHWH

The significance of the Movements of the Chariot of YHWH depends on the interpretation of the meaning of the Chariot structure, as a whole. Firstly, the question may be asked, why does the Chariot move at all? Are the movements meant to have material or metaphorical meaning or both?[782] In the discussion about the Living Creatures (see 'Middle Tier') it was shown that the movements of the Chariot signify a metaphorical movement of man 'progressing through the world' or confronting new situations, and so can relate the innate psychological characteristics of the Living Creatures to the way humans make moral decisions. This would fit with Lefebvre's 'mental' space.[783]

Secondly, the sudden lightning-like movements of the chariot seem to be telling us something more specifically about sacred space: the movements are initiated by the spirit and are accompanied by a lightning flash. Each movement appears to be a specific spiritual decision.

Thirdly, there are two key movements that the Chariot makes (Appendix 9: Movements of the Chariot of YHWH). The first is in Ez 11.23 when the Chariot goes up from the midst of the city and stands on the mountain to the east of the city. The second is in Ez 43.2 when it comes back into the temple from the east gate many years later. These movements have their own spatial story to tell. Knott divides her analysis of religion and space into four areas: space and the sacred, geography of religion, religion and globalisation and religion and locality.[784] In terms of this division, these two key movements the Chariot

---

making in the world. The double metaphor hides a powerful insight into crucial forces that protect our decision making regarding the sacred. This fits very much with Lakoff and Johnson's linking of our cognitive processes and reason and the way in which this interpretation of the Chariot movements uses orientational metaphor. Lakoff, G. & Johnson, M. (1999), *Philosophy in the Flesh*, New York: Perseus, p. 148.

782 Knott, K. (2005), *The Location of religion: A Spatial Analysis*, London: Routledge, p. 12.

783 Lefebvre, H. (1991), *The Production of Space*, Oxford: Blackwell, p. 299.

784 Knott, K. (2005), *The Location of religion: A Spatial Analysis*, London, Routledge, p. 94

makes, seem to be telling us about space and the sacred, and the lightning movements possibly refer to 'globalisation'.

Fourthly, there is the importance of Ezekiel encountering the Vision in a specific geo-graphic location (Knott's religion and locality). The geographic location seems to 'emphasizes divine mobility over the notion thatYHWH could only be encountered in the Jerusalem temple.'[785] The location of the initial Chariot Vision may therefore be just as important to establishing a new idea of 'divine mobility' as are the actual chariot movements recorded in Ezekiel. Greenberg says that the location of the initial Chariot Vision was not to prove that prophecy could occur in a foreign, 'impure' land. The exiles were confident of God's attention,[786] as prophets were among them (Ez 13:9).

As discussed above there are two main interpretations of the Chariot structure. Firstly, and as declared in the text of Ezekiel, it is a manifestation of the presence and glory of God (this leads to the interpretation which I will call the Idealised Temple or IT interpretation). Secondly if the 'likeness of a man' seated on the top of the Chariot Vision is interpreted as an analogy of Christ, and other features in the Chariot are interpreted as pointing to the Trinity (see 'Trinitarian Signs in the Chariot Vision') the Chariot Vision movements can be similarly interpreted (this will be called the Christocentric or CC interpretation). These interpretations are not mutually exclusive, but the CC interpretation is a particular interpretation of the IT interpretation. Because movements are a major feature of the Chariot Vision, both interpretations involve a consideration of Sacred Space.[787] These two interpretations will now be considered in detail.

---

[785] Odell, S. (2005), *Ezekiel*, Smyth & Helwys Bible Commentary, Macon: Smyth & Helwys, p. 16.

[786] Greenberg, M. (1983), *Ezekiel 1–20*, New York: Doubleday, p. 59.

[787] Lévi-Strauss, C. (1962), *La Pensée Sauvage*, Paris: Plon, p. 17.

The Idealised Temple Interpretation

The two key movements of the Chariot Vision mentioned above are central to the IT interpretation. In the first of these two key chariot movements, the Glory of God moves out of the Temple and the city and is seen to stand over the mountain to the east of the city[788] (Ez 11:23). This departure comes after the city of Jerusalem was defiled by the slaughter of those who committed abominations, beginning at the sanctuary (Vision of the Executioners Ez 9:1-11), because 'the guilt of the house of Israel and of Judah is exceedingly great' Ez 9:9. The abominations committed appear to be the cause of its departure. Following the slaughter, fire was taken out from between the 'cherubim' of the chariot to be scattered over the city. In Ez 11.17 before the Glory of God leaves, God promises to gather the exiles from where they have been scattered and to put a new spirit and a heart of flesh in them. The Exile caused intense theological shock, but the people maintained their confidence in YHWH's obligation to rescue them. This sense of security was based on the conviction of an inseparable bond between YHWH, the territory (Land of Canaan) and the people (nation of Israel).[789] (Figure 24 after Block[790])

When YHWH established a covenant with Israel, Ezekiel puts honouring the name of YHWH as a principal charge of Israel.[791] In rebelling against God, the name of YHWH is defiled, and the blessings of the covenant are withheld. Part of the charge of the covenant was for Israel to honour the name of YHWH 'in the centre of the nations with countries all around her' (Ez 5.5). In rebelling against God, Israel's central geographical position of witness to YHWH in relation to her neighbours is threatened. The movement of the Chariot out of the Temple, (as well as the location of the first appearance of the Chariot Vision

---

[788] The mountain on the East side of the city is the Mount of Olives (*har hazzētîm* הַר הַזֵּיתִים).
[789] Block, D. I. (1997), *The Book of Ezekiel Chapters 1–24*, Grand Rapids: Eerdmans, p. 52.
[790] Block, D. I. (1997), *Ibid.*, p. 7.
[791] Block, D. I. (1997), *Ibid.*, p. 53.

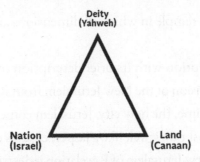

Figure 24: The inseparable bond between YHWH, the territory (Land of Canaan) and the people (nation of Israel).

in Babylonia rather than in Jerusalem), may be a forewarning that there will no longer be one sacred place, the Temple, but now wherever the Hebrew people meet to pray becomes sacred following the destruction of the Temple. Ezekiel saw the vision of the chariot moving out of the Temple (Ez 11:23) in about 592 BC[792] after the first siege of Jerusalem (598 BC) but before the destruction of the Temple in Jerusalem (587–6 BC). He saw the Idealised Temple Vision in 573 BC, thirteen years after the destruction of the Jerusalem Temple and three years before his death. The return of the Vision of the Chariot of YHWH within the Idealised Temple Vision (Ez 43.2) links these two visions. Since there is a movement of the Chariot of YHWH into the temple, the Idealised Temple Vision becomes central to the interpretation of the movements of the Chariot. The Idealised Temple Vision has an eschatological flavour,[793] particularly since the temple described and measured in the vision differed from both the first and second Temples in Jerusalem and therefore suggested an 'idealised' temple that was yet to be built. The temple description, however, is not presented as a blueprint for a temple to be constructed with human hands.[794] The Idealised

---

[792] Block, D. I. (1997), *The Book of Ezekiel Chapters 1–24*, Grand Rapids: Eerdmans. p. 29.

[793] Block, D. I. (1998), *The Book of Ezekiel Chapters 25–48*, Grand Rapids: Eerdmans, p. 504.

[794] Block, D. I. (1997). *Ibid.*, p. 59.

Temple Vision is of a temple in which all dimensions are carefully prescribed by God.

This temple description with its brief description of a new Jerusalem (Ez 48:30-35), recalls the vision of the New Jerusalem from Revelation, (Re 21:11) in which, at the end of time, the holy city, Jerusalem comes down out of heaven, 'its radiance like a most rare jewel, like a jasper, clear as crystal'. Like Ezekiel's new Jerusalem, the New Jerusalem of Revelation is described as square in shape. and has twelve gates named after the twelve tribes of Israel. The measurements of these two cities are vastly different, however, with Revelations (Re 21:16), giving the length of one side of 12,000 stadia, which is about 2,280 kilometres.[795] The height of the walls are 66 meters (144 cubits). The measurements of Ezekiel's Jerusalem is a more rational, 4,500 cubits (2.05 kilometers).

Block would see both of these descriptions as different restatements of the covenantal relationship between YHWH and the people of Israel, the careful reassurance of the gift of land as a demonstration of the durability of the covenant for all time.[796] There are also similarities with Zechariah's third vision of 'the man with a measuring line' (Ze 2:2). Here the man with a measuring line is a surveyor who is ready to start measuring off the breadth and length of Jerusalem. Zechariah's new Jerusalem however would have no walls. The population would be abundant and prosperous and the Lord will come and dwell in the city to protect it.[797]

The Idealised Temple Vision represents: (a) idealised perfection, (b) the eschatological future c) of spiritual rather than scientific significance. The significance of the return of the Glory of God to the temple appears to suggest that in the eschatological future, God will return to a spiritually perfected people rather than to a real building of a third Temple. In this final vision

---

[795] A biblical 'stadium' is about 190 metres.

[796] Block, D. I. (1998), *Ibid.*, p. 354.

[797] Cody, A. (1990), 'Zechariah,' *New Jerome Biblical Commentary*, London: Burns & Oates, p. 354.

of Ezekiel, where God is, *there* will be Zion.[798] Smith would put this as a transfer of the sacred from the Temple, to inner, contemplative mental space.[799] After the Exile, the cherubim throne of YHWH in the Temple in Jerusalem, which was regarded as the footstool of YHWH, became an abstract symbol and 'all Jerusalem' became the throne of YHWH.[800] This interpretation fits into Knott's 'Space and the sacred' category.[801]

The Christocentric Interpretation

Origen was the first to point out the link between Christ and Ezekiel.[802] Origen insisted that all of Scripture be interpreted in the light of Christ. He held that this was particularly true of Ezekiel, and he showed several parallels between Christ and Ezekiel. Like Christ, Ezekiel began his ministry in his thirtieth year. He also saw a parallel between Ez 1:1 'the heavens were opened' and Lk 3:21 'the heaven was opened' at the baptism of our Lord. Ezekiel was on the banks of the River Chebar as was Christ at the Jordan, (Lk 3:23). The Holy Spirit descended as a dove and in Ezekiel the fire and *hashmal* in the centre of the cloud introduced the Chariot of YHWH when it appeared. Origen's view that Christ as the key to Scripture is particularly true of Christian exploration of the Chariot Vision.[803] Several of the Church fathers see Christ in the Chariot Vision as reviewed in the section 'Church Fathers interpretation of the Chariot Vision'.

---

[798] Block, D. I. (1997), *The Book of Ezekiel Chapters 1–24*, Grand Rapids: Eerdmans, p. 59.

[799] Smith, J. Z. (1987), *To Take Place: Towards Theory in Ritual*, London: University of Chicago Press, p. 117.

[800] Weinfeld, M. (1976), 'Jeremiah and the Spiritual Metamorphosis of Israel,' *ZAW*, **88** (1), pp. 2–55; p. 20.

[801] Knott, K. (2005), *The Location of Religion: A Spatial Analysis*, London: Routledge, p. 94.

[802] Christman, A. R. (2005), *'What did Ezekiel See?,'* Leiden: Brill, p. 29.

[803] Christman, A. R. (2005), *Ibid.*, p. 29.

Block is the principal modern author who identifies the descent of the Chariot Vision with Christ, the Messiah; descending and dwelling among us.[804] He says that overt references to the Messiah are few, but there is an allusion to a *rak* or shoot in the topmost crown of the cedar in the epilogue of the fable of the Eagle and the Vine (Ez 17:22).[805] He sees links to Christ as the Messiah in several parts of Ezekiel however, including the Vision of Water Flowing from the Temple (Ez 47);[806] the Oracle against Gog (Ez 38) (through the book of Revelation especially, Re 19:17–21);[807] the Vision of the Dry Bones (Ez 37);[808] The Shepherd of Israel metaphorical narrative (Ez 34);[809] The Eagle and the Vine extended metaphor (Ez 17:22-24)[810] and the Vision of the Executioners (the scribe with the inkhorn[811] and the forehead signing with Taw[812]) (Ez 9:4). As discussed below (under 'Trinitarian Signs in the Chariot Vision'), Block identifies the 'spirit' of the Chariot Vision with the Holy Spirit.[813] In addition there are several other potential links with Christ:

1. The 'likeness of a man' seated on the throne on the top of the Chariot Vision has been identified with Christ, by several Church Fathers, including Eusebius, Theodoret of Cyrus, and Gregory the Great.[814]

2. The Mount of Olives: when the Chariot of YHWH leaves the Temple in Ez 11:22, it moves to a position over the Mount of Olives after leaving the

---

804 Block, D. I. (1998), *The Book of Ezekiel Chapters 25–48*, Grand Rapids: Eerdmans, p. 746.

805 Block, D. I. (1997), *The Book of Ezekiel Chapters 1–24*, Grand Rapids: Eerdmans, p. 57.

806 Block, D. I. (1998), *The Book of Ezekiel Chapters 25–48*, Grand Rapids: Eerdmans, p. 702.

807 Block, D. I. (1998), *Ibid.*, p. 493.

808 Block, D. I. (1998), *Ibid.*, p. 392.

809 Block, D. I. (1998), *Ibid.*, p. 308.

810 Block, D. I. (1997), *Ibid.*, p. 553.

811 Block, D. I. (1997) *Ibid.*, p. 305.

812 Block, D. I .(1997) *Ibid.*, p. 311.

813 Block, D. I. (2013), *By the River Chebar: History, Literary, and Theological Studies in the Book of Ezekiel*, Eugene: Cascade, p. 157.

814 Christman, A. R. (2005), *'What did Ezekiel See?,'* Leiden: Brill, p. 30.

Temple. At this point the Temple had been profaned, and fire scattered over the city. In the New Testament, Christ took the sins of mankind on himself when he prayed on the Mount of Olives before his Passion. (Lk 22:39) The Temple had also recently been profaned by those buying and selling (Mt 21:12) and in his anger, Jesus cleared the Temple.

3. An argument is made above that the coals between the Living Creatures could be an analogy of a sacrificial holocaust and the fire emanating and scattered over Jerusalem could be interpreted as the blood of the sacrifice (See 'Spirit in Ezekiel and the Chariot'). The supreme sacrifice that took place in Jerusalem was the crucifixion and death of Christ. In a Christocentric interpretation, the coals and fire of the Chariot Vision can therefore be interpreted as the sacrifice of Christ at Golgotha. In this interpretation, the fire that was spread around the city in Ez 10:7, is the blood shed by Christ in and around the city of Jerusalem, and a foreshadowing of Christ's redemptive sacrifice.

4. The Hebrew word *galgal* may have special significance in the CC interpretation. A discussion above shows how *galgal* comes from the same Hebrew root (*galal*) as Galilee, where Christ was born and centred his ministry. The word Golgotha, the place where Christ was crucified and atoned for mankind's sins, also has the same root. The word *galgal* can be seen to encompass the places of both Christ's mission and his death. As noted above, in Ezekiel, *galgal* is particularly used in relation to the fire between the wheels (Ez 10:2, 6; see Appendix 2), which supports the link between the fire, the coals, sacrifice and Golgotha.

5. The dual nature of the wheel both as a concrete structure and a rapid whirling movement would fit with the dual nature of Christ who was both man and God. The rapid lightening-like movements of the Chariot can be interpreted as the spread of the gospel. Finally, if the return of

the Chariot of YHWH to the temple, in the Idealised Temple Vision represents the return of Christ, this supports the eschatological character of the Idealised Temple Vision. If Christ is returning at the end of time, then the temple would represent his perfected, faithful Church.

6. The wheel within a wheel (Ez 1:16) was interpreted by several of the early Fathers as the New and Old Testaments, with Gregory stating that the New Testament ran within the Old.[815] Bonaventure has said that that mathematics is numbers and figures and mathematics leads to Christ:[816] The perfect figure is a circle, with its infinite circumference and infinitesimal centre. The inner circle could therefore represent Christ and the Eucharistic host.[817] In this context, the rigid structure of the wheels recalls the rigid structure of Christ's cross.[818]

7. Ryan sees a link between the departure of the glory of God from the Mount of Olives in Ez 11:22 and the Ascension of Christ (Ac 1:9) from just outside the village of Bethany (Lk 24:50) which is on the Mount of Olives.[819]

In a Christocentric interpretation, the arrival of the Chariot Vision on earth can be seen to foreshadow the Incarnation, the coming of Christ to earth.[820] The departure of the Chariot to the East of Jerusalem would be a foreshadowing

---

[815] Gregory the Great, *Homilies on Ezekiel*, I.6.12.

[816] Bonaventure (1956), *The Journey of the Mind to God*, Indianapolis: Hackett, p. 22.

[817] Because of its symmetry the circle is considered as the perfect shape. It is the symbol for the total symmetry of the divine. The Greek scholar Proclus (5th century AD) wrote: "the circle is the first, the simplest and most perfect form". As Christian symbol it represents eternity, and the sleeping eye of God (Pr 8:27: uses *hûg* (חוג) for circle). Circle at *http://www.2dcurves.com/conicsection/conicsectionc.html* accessed 15/3/20.

[818] If *wîdêhem* in Ez 10:12 is interpreted as 'spokes', then spokes extending from the axis of the wheel make the shape of a cross (see 'The Wheels as a Class of Angels').

[819] Ryan C. L. (2014), *Concerning Ministerial Visions: An Introduction to Ezekiel's Visions*, Vol. 2, Birmingham: Capital Cube Creatives, p. 72.

[820] Block, D. I. (1998), *The Book of Ezekiel Chapters 25–48*, Grand Rapids: Eerdmans, p. 746.

of the Ascension. The temple of the IT Vision would be synonymous with the spiritual perfection of the people and the return of Christ at the end of time, (since the New Jerusalem of Revelation (Re 21:11) has no temple). This interpretation would also fit with Knott's 'Space and the Sacred category'.[821]

The Rapid Movements of the Chariot

In the context of the idealised temple interpretation, the lightning-like movements of the Chariot in Ez 1.14 could relate to the way the people of Israel had been scattered and that the time was coming after the destruction of the second Temple (in 70 AD), when there would be a proliferation of sacred spaces throughout the world for the people of Israel, (Knott's 'globalisation' category).

In the context of the Christocentric interpretation, a possible interpretation is that the rapid movements represent the multiple sacred spaces that were set up through Christ's presence on earth. The movement of the wheels may represent Christ's presence on earth, at the time of his ministry, from Galilee to his death on Golgotha.

If the coals and the fire are taken as the sacrifice of Christ's death on Golgotha, and the Living Creatures as priests, then the fire and coals could represent the sacrifice of the mass. The movements of the Chariot then represent the profusion of tabernacles ('a little sanctuary'[822] *miqdāš mĕ'aṭ*, Ez 11:16) and the spread of the sacrifice of Golgotha throughout the world. In addition, if as mentioned above, the inner wheel of the 'wheel within the wheel' of the Chariot of YHWH (Ez 1.16) is an analogy for Christ,[823] the rapid lightning

---

[821] Knott, K. (2005), *The Location of religion: A Spatial Analysis,* London: Routledge, p. 94

[822] Block, D. I. (1998), *Ibid.,* p.358.

[823] Gregory the Great said that the wheel in the midst of a wheel is 'the New Testament encompassed by the Old'. Saint Gregory the Great (1990), *Homilies on the Book of the Prophet Ezekiel,* Etna: Center for Traditionalist Orthodox Studies.

movements of the wheels would also signify the profusion of sacred spaces brought about through Christianity.[824]

Conclusion of the Chapter

The seasonal timing, celestial orientation and solar orientation of the principal Chariot Vision movements appear to relate to the mood of the relevant Ezekiel texts. Return of the chariot to the Idealised Temple could represent God's return to a spiritually perfected people in the eschatological future. A Christ-centred interpretation is also a good fit with the chariot movements.

---

[824] It was argued above that the Hebrew words *bārāq* and *bāzāq* which are usually interpreted as 'lightning' could in fact be electrostatic discharges. The rapid movements
then could be interpreted as dissemination of sacred space through electronic media.

Chapter 11

The Interaction of Reason and Spirit

Overview of the Chapter

This chapter reviews the instances of interaction of reason and the spirit in the three levels of the Chariot Vision. It also explores how the spirit may input into science and scientific reasoning.

The Pythagorean Philolaus suggested in the fifth century BC, that reason was separate from perception.[825] In the Semitic world, the Hebrew word (*lēb*) is used for both mind and heart and tends to therefore integrate the activities of reason and feeling.[826] However most of its uses in the Old Testament refer to intellectual, rational functions.[827] The separation between mind and body that these two forces represented came to its peak in Descartes 'I think therefore I am', but as intuited by the Hebrews, rationality and emotions are closely linked. The flaw that Israel's Babylonian exiles had developed was that they

---

[825] Diels-Kranz 44, B13, 1–6, in: *Theology of Arithmetic* (1922), Falco trans., Leipzig: Teubner, pp. 25, 17; Launderville D. F. (2007), *Spirit and Reason,* Waco: Baylor University Press, p. 53.

[826] Launderville, D. F. (2007), *Ibid.,* p. 10.

[827] Wolff, H. W. (1974b), *Anthropology of the Old Testament,* Philadelphia: Fortress, p. 46.

had divorced reason from emotion.[828] The interface between rationality and emotion is central to the message of Ezekiel. The balancing of rationality and emotion is achieved in the heart[829] but correct balancing of reason and affect requires the input of the spirit.[830] The spirit therefore is essential for right reason.

Modern psychology research shows that reasoning requires both rationality and emotion with emotion playing a cognitive guidance role in decision making.[831] The brain stores many different personally categorised scenarios ('contingencies') which it uses for reaching goals in specific time-frames.[832] The mental balancing of these contingencies in making personal decisions is mostly subconscious. The activity of the heart is also subconscious, although in critical emotional events in decision-making, the heart-rate may increase, and a person may consciously feel this increase in heart rate.[833] Complex decision making is simplified by what have been called somatic markers (or unpleasant 'gut' feelings) which highlight options that should be rapidly eliminated from consideration. These feelings, emotions or desires are accompanied by recordable changes in the body, (such as skin conductance – as in a lie-detector test), that are mediated by the autonomic nervous system.[834] These feelings equate with

---

828   Launderville, D. F. (2007), *Ibid.*, p. 45.

829   Pinckaers, S. (1995), *The Foundations of Christian Ethics,* Washington DC: The Catholic University of America Press, p. 42.

830   Launderville, D. F. (2007), *Ibid.*, p. 348.

831   Damasio, A.R. (2005), *Descartes' Error: Emotion, Reason and the Human Brain,* New York: Harper Collins, p. 130.

832   Damasio, A.R. (2005), *Ibid.*, p. 183.

833   Damasio, A.R. (2005), *Ibid.*, p. 254; Fuster, J. M. (1985), 'The prefrontal cortex, mediator of cross-temporal contingencies,' *Hum Neurobiol,* 4 (3), pp. 169–79.

834   Damasio, A. R. (2005), *Ibid.*, p. 174; The autonomic nervous system is responsible for a racing heartbeat in response to an emotional situation. It acts largely unconsciously to regulate bodily functions. This system is the primary mechanism in control of the fight-or-flight response and responds to emotional cues.

the 'passions' which St Thomas Aquinas said can make a positive contribution to moral action under the guidance of the virtues.[835]

To have a heart in the right place was essential for Ezekiel's prophetic role.[836] In this context the crucial phrase from Ez 36:26 is 'and I will take away from you your heart of stone and give you a heart of flesh' *wahăsirotî lēb hā'eben mibběsarām wěnatattî lāhem lēb bāśār.* Here YHWH intimates that the Israelites' idolatrous ways are due to a disconnection of their 'hearts' from their decision-making process. The 'heart' of flesh is a renewed locus for their noetic and affective capacities and incorporates a restored harmonious relationship between thinking and feeling. [837]

In Ezekiel, the heart is closely connected with the spirit. Three times YHWH promises to give 'a new heart and a new spirit' (Ez 11:19; 18:31; 36:26) to the Israelites and in two of these mentions, he also links giving of the spirit to taking away their heart of stone and giving them a heart of flesh (Ez 11:19; 36:26). The spirit is essential for correct functioning of the heart. A hard-hearted person *ḥizqê-lēb* (Ez 2:4) is unable to read sensory information properly. The spirit is intimately linked to the rationality of the heart and the spirit has an empowering, guiding and integrating effect.[838] Launderville sees a connection between humans and the cosmic order through the spirit that enables the right ordering of perception and allows a 'critical' rationality (*nous*) that apprehends truth directly or intuitively and not by a process of reasoning.[839]

---

[835] Pinckaers, S. (1995), *The Foundations of Christian Ethics,* Washington DC: The Catholic University of America Press, p. 225.

[836] Launderville, D. F. (2007), *Spirit and Reason,* Waco: Baylor University Press, p. 232.

[837] Launderville, D. F. (2007), *Ibid.,* p. 348.

[838] Launderville, D. F. (2007), *Ibid.,* p. 55.

[839] Launderville, D. F. (2007), *Ibid.,* p. 93.

Living Symbolism of the Israelites

An intriguing living form of symbolism is put forward by Launderville, who says that Ezekiel fosters a type of symbolic thinking that made the Israelites living symbols of YHWH. When they were in Egypt, YHWH chose the Israelites to be symbols of his presence in the world. The Israelites had developed a disordered way of perceiving reality.[840] Ezekiel's main task was to restore the interaction of reason and spirit in the hearts of the Israelites by sensitizing their hearts to the emotional cues of his truth. Their acceptance of a new heart was the first step to their renewal as symbols of YHWH. The second step was the giving of a new spirit which would be a participation in YHWH's Spirit.

Integration of Reason and Spirit in the Chariot of YHWH

The 'Likeness of a man': Symbol of balancing of noetic and affective forces

The discussion above under 'Upper Tier' puts forward the appearance *hašmal* above the loins of the 'likeness of a man' as representing the heart and the gold and silver mixture of electrum potentially standing for the interaction of reason and emotion. The Spirit may also be represented as the fire surrounding the *hašmal* as this is the first part of the Chariot to appear in the storm cloud in the original Chariot Vision (Ez 1:4). The *hašmal* itself can be interpreted as a symbol of the 'heart of flesh' (Ez 11:19) the Israelites will be given with the 'new spirit' as the surrounding fire. The *hašmal* and fire of the likeness of a man would be a symbol of correctly integrated reason and spirit. This image in the 'likeness of a man' would in this way visually encapsulate the giving of a new heart and a new spirit to the Israelites (Ez 11:19; 18:31; 36:26).

---

[840] Launderville, D. F. (2007), *Ibid.*, p. 348.

The Living Creatures: Reason and Spirit

The 'Middle Tier' chapter on the Living Creatures shows that the connection between reason and emotion is part of their physiognomy. In the discussion on the Living Creatures, it was mentioned that they have 'spiritually-dependent directional potency': they have their own will and decide their direction, but they are dependent on the spirit to achieve the correct direction. The Creatures are an analogy for the incorporation of the spirit into decision making. That chapter hypothesised that right reason is analogised in the Creatures by two axes: cognitive (or thinking) and affective (or feeling). The right balance between these that achieves right reason, is brought about when the Creatures become docile to the spirit. Since the Creatures balance rational and affective capacities[841] they are also a model for the balancing of decisions by the 'heart'.

For the Creatures, becoming docile to the spirit is the equivalent of having a new heart. It is a removal of pride from the application of the will and a focusing of the will on the Spirit of YHWH. The continuing docility of the Creatures to the spirit allows them to have a spirit-directed relationship between themselves and with YHWH. Like the Israelites, they have become symbols of YHWH's presence.[842] Their thinking is therefore of a symbolic kind as it points beyond themselves to something else and they participate in the reality to which they point.[843]

---

[841] Launderville, D. F. (2007), *Spirit and Reason*, Waco: Baylor University Press, p. 348; The eagle's intuition gives the human-eagle cognitive axis 'noetic' or intuitive reasoning capacity. The bull and lion make up the emotional axis. Intuitive apprehension of the truth is balanced against emotion in the 'heart' and is important for compassionate action: McCraty R, Zayas M. (2014), 'Intuitive Intelligence, Self-regulation and Lifting Consciousness,' *Glob Adv Health Med*, **3** (2), pp. 56–65.

[842] Launderville, D. F. (2007), *Spirit and Reason*, Waco: Baylor University Press, p. 348.

[843] Launderville, D. F. (2007), *Ibid.*, p. 7.

The Chariot Wheels: A sign of rationality

As discussed above, the wheels can be interpreted as a metaphor for inferior reason. The wheels of the chariot in comparison to the Living Creatures are much more in keeping with 'signs',[844] rather than symbols, as their distance and direction in relation to the earth is measured univocally by measurement or compass orientation. The wheels' spiritual docility is accessed through the Creatures, and the interplay between emotion and cognition has already been resolved in the Creatures. The spirit is however essential for the correct functioning of the 'rationality' of the wheels and inferior reason.

Science and Intuition: Structuring of Emotion and the Spirit

Science is a discipline whose primary object is the understanding of the spatial-temporal world. Ideas are tested and accepted based on evidence generated through scientific method. The great power of science lies in the ability to relate cause and effect.[845] Scientific truths are universal and are typically independent of time, place and circumstance,[846] and every scientific statement must be testable.[847] The definition of science used in this thesis is that there must be evidence of hypothesis and experiment.[848] That is, there has to be the conceiving of a new idea and it has to be examined logically.[849] It was discussed above that empirical science is encompassed by inferior reason and since the spirit is necessary for the correct functioning of the wheels and inferior reason, the spirit is also necessary for empirical science.

---

[844] Launderville, D. F. (2007), *Ibid.*, p. 6.

[845] Crutchfield, J. P. *et al.*, (1986), 'Chaos,' *Scientific American*, **54** (12), pp. 46–57.

[846] Wartofsky, M. W. (1968), *Conceptual Foundations of Scientific Thought*, New York: Collier-Macmillan, p. 23.

[847] Popper, K. (1972), *The Logic of Scientific Discovery*, London: Hutchinson, p. 48.

[848] Wartofsky, M. W. (1968), *Ibid.*, p. 181; 'Experiment' means experimental methodology, particularly through measurement, to reach a conclusion by inductive reasoning.

[849] Popper, K. (1972), *Ibid.*, p. 31.

Scientific discovery is not a purely logical process and every discovery contains an irrational or intuitive element.[850] Albert Einstein said:

> (universal laws)... can only be reached by intuition, based upon something like an intellectual love (*Einfühlung*) of the objects of experience.[851]

Intuition is a cognitive process that is however subconscious and is feelings-based, and so linked to emotion. Not only is intuition central in rational decision making, but emotion also influences decision making.[852] Intuition, however, must be kept separate from emotion and the correct structuring of emotions[853] is essential for right reasoning. This correct balancing of reason and affect requires the input of the spirit.[854] The spirit therefore is essential at two distinct points in empirical science: firstly, at the stage of formulation of scientific questions, to correctly balance intuition, and secondly at the stage of logical scientific reasoning (inferior reason) in order that analyses and conclusions may conform to 'right reason'.

The three tiers of the Chariot Vision are therefore three different 'spheres' of interaction between the rationality and the spirit: The upper tier contains a symbol of YHWH's promise to the Israelites to make them living symbols of himself. The middle tier is an analogy for how the Israelites should make

---

850   Popper, K. (1972), *Ibid.*, p. 32.

851   Einstein, A. (1934), 'Address on Max Planck's 60th birthday,' in Einstein, A. *Mein Weltbild*, p. 168. English translation by Harris, A. (1979), *The World as I see it*, New York: Philosophical Library, p. 125.

852   Strong emotions can affect intuition: Immordino-Yang, M. H., Damasio, A. (2007), 'We Feel, Therefore We Learn: The Relevance of Affective and Social Neuroscience to Education,' *Mind Brain and Education* 1 (1), p. 3; Zheng, Y. *et al.* (2017), 'The Influence of Emotion on Fairness-Related Decision Making: A Critical Review of Theories and Evidence.' *Front. Psychol.*, 8, article 1592.

853   Pinckaers, S. (1995), *The Sources of Christian Ethics*, Washington DC: The Catholic University of America Press, p. 42: The virtue of prudence is important in this respect.

854   Launderville, D. F. (2007), *Spirit and Reason*, Waco: Baylor University Press, p. 348.

spiritually-guided rational decisions. The lower tier is a sign that even in 'inferior reason' such as measurement, that spiritual guidance is also necessary. And that science also entails an interaction between reason and spirit.

Conclusion of the Chapter

The three tiers of the Chariot Vision represent three separate spheres of interaction of reason and spirit. Scientific reasoning entails an interaction of reason and spirit.

Chapter 12

Rationality, Science and Revelation

Overview of the Chapter

The links of between rationality and revelation in the Chariot Vision and how Maimonides' divine science and modern science are exemplified are explored.

The 'science' of Maimonides' divine science is not identical to modern empirical science which is a 'natural' science.[855] Maimonides implies that the Chariot Vision reveals *divine* information in a rational way. The arrangement of the tiered structure of the Vision can be interpreted as 'science', as it revels in a rational way. The chariot structure links the 'science' of Maimonides and empirical science as the chapter 'Hierarchies and Fuzzy Epistemology' shows that the three levels of the Chariot are congruous with a three-level science-based epistemology[856] as applied to the movements of the Chariot. The FE structure shows a progressive revelation of the action of the spirit on movement of the Chariot: spirit > lightning > movement. The chapter also shows an

---

[855] Kreisel, H. (2015), *Judaism as Philosophy: Studies in Maimonides and the medieval Jewish Philosophers of Provence*, Boston: Academic Studies Press, p. 252.

[856] Seising, R. (2008), 'Fuzzy epistemology: The fuzziness of Experimental Systems,' available from: *https://www.researchgate.net/publication/4338088_Fuzzy_epistemology_The_fuzziness_of_experimental_systems* [accessed 36/9/2021].

epistemological equivalence across the three concrete objects (the wheels, the measuring line and the scales) of the three visions analysed (Figures 21, 22 and 23). This suggests a similar science-based way of gaining knowledge for all three concrete objects.[857]

The Chariot firmly links revelation with rationality: At the upper level, the appeal to the heart targets a key flexure point in decision making between intuition and emotion, which gives a rational means of sensitizing decision making. At the middle level, the ability to make correct moral decisions is seen to be a rational balance between character-forces inherent in humans, incorporating reason and emotion, and not an arbitrary pursuit of rules. The lower level shows that natural phenomena should be understood with the help of measurement[858] and empirical science. In addition to this, as discussed above, the three levels of the Chariot are revelatory in different ways: The Upper Tier is a sign of covenantal love, as well as a symbol for restoration of the Israelites.[859] The Middle Tier reveals the psychological forces underlying right reason, and the Lower Tier is a sign of the importance of measurement and 'science' in revelation. These three levels themselves are rationally related to each other.

---

[857] This also supports that within the context of Ezekiel, the wheels may have a metrological function.

[858] Measurement has a pivotal place in empirical science, so much so that many would say that without measurement of some kind, you do not have true science. Lord Kelvin (Thomson, W. (Lord Kelvin) (1883), lecture on 'Electrical Units of Measurement' to the Institution of Civil Engineers on 3 May 1883) famously stated:

> ...when you can measure what you are speaking about, and express it in numbers, you know something about it; but when you cannot measure it, when you cannot express it in numbers, your knowledge is of a meagre and unsatisfactory kind: it may be the beginning of knowledge, but you have scarcely, in your thoughts, advanced to the stage of science.

[859] Launderville, D. L. (2007), *Spirit and Reason*, Waco: Baylor University Press, p. 348.

By the same token, the Chariot is linked to science as a revelatory mechanism, with the wheel underscoring science and technology and the Living Creatures' decision-making process concurring with modern psychological dynamics. There is therefore prima facie evidence of 'science' within the Chariot Vision. Within the text of Ezekiel too, there is a 'scientific experiment' which shows a three-level FE science-based structuring and also evidence of inductive reasoning. The comparison of the Vision of Water flowing from the Temple with Archimedes experiment, shows that it fulfils basic criteria for a scientific experiment and it can be interpreted as an indication to Ezekiel to use measurement (and hence science) in order to understand nature. This supports the presence of scientific reasoning in Ezekiel. Additionally, the three Bible passages analysed show that the FE structure is present in different parts of the Bible.

Fuzzy Epistemology within Bible Structures

Having shown that the three layers of FE are discernible in Ezekiel a question that arises is, whether the objects of a three-level experimental system (epistemic/theoretical; fuzzy; real: see Figure 19), are the same as the three levels that are discernible in the Biblical passages above. Following the tenets of scientific realism, the concrete object of the lowest layer of the experimental system has mind-independent or objective value and represents real knowledge. Concrete objects in the Bible, inasmuch as they are divinely revealed and linked to a message or command from God, can be taken to be real knowledge as well. The difference between the two is that in the first case the knowledge has been gained through empiric scientific experimentation, and in the second, by revelation. In the second case the message from God is what is important and is revealed by tracing a link from the concrete object. In the first case, the object itself is the new knowledge, and occurs by 'defuzzification' of the fuzzy or middle layer of the experiment.

In hierarchies in the Bible, if there is a genuine link between a 'fuzzy' object like a bolt of lightning and a concrete object such as an arrow, [860] then it could reasonably be said that the object in the real layer may be present in the middle layer in a 'fuzzy form'. In the Chariot of YHWH example, the text of Ezekiel clearly links the spirit, lightning flashes and wheels through the movements of the Chariot. Therefore, the epistemic reason for the Chariot wheel movements (which might be, for example, 'to demonstrate the movements of the Hebrew people') is contained within 'the spirit' (upper level) which commands the movements. This 'epistemic object' is represented in the middle level by lightning flashes and then as moving wheels in the technical or real layer. The lightning flashes thus can be said to represent a 'fuzzified' form of the wheel movements. The middle layers of both the Experimental System and Biblical revelation are therefore compatible with 'fuzzified' knowledge. The Experimental and Biblical structures have similar levels therefore and can be seen to proceed from upper to lower levels in a similar fuzzification/defuzzification sequence.

A further example can be seen in the Living Creatures, for which, several different interpretations have been suggested above (See 'Living Creatures Significance'). Is only one of these correct? A fuzzy epistemological interpretation would be that the four Living Creatures represent the fuzzy level of a divine message. Since the data 'contained in' the four Creatures is fuzzy it can be interpreted in different ways, and there is not just one correct answer. Different interpretations can be opened up from tracing back from a real object. For example, the 'straight leggedness' of the Creatures suggests the possibility the Creatures represent the priesthood, which seems to find a reasonable fit. The wheels suggest 'science', in which the character-forces of the Creatures become a model for scientific decision making as noted in the previous section above. The multiple interpretations possible for the Creatures suggests that other

---

[860] Ps 144:6: 'Flash your *lightnings*; rout the foe, shoot your *arrows* and put them to flight' – (*lightnings* can be considered a 'fuzzy' object and *arrows,* the concrete object).

Biblical passages, for example those containing a symbol, may also be opened up to interpretation in the same way.

Why would the layers of a scientific Experimental System be detectable within the Bible? Since, from the foregoing, it does not appear that the similarity between the experimental system and the Biblical structures is merely due to chance, it does seem that there is an embedding of the three-level FE structure within certain Biblical passages. Since the Bible predates the development of scientific experimental systems, it suggests that the experimental system structure is one way in which God interacts with humanity in the Scriptures. In this interpretation, God uses a two-step method of revelation: firstly, a fuzzified object which may be a natural phenomenon or speech, and second, a measurable concrete delineation of the revelation. The fact that the human brain uses fuzzy logic to reason, means that this is a two-pronged way of communication: first directly to our senses through a 'fuzzy' object which is incompletely understandable, and then secondly to our rational logic by using measurable ('technical') objects (see Ps 144 in preceding footnote). The two ways of communication support each other and are scientifically linked. 'Science' may therefore be part of the way in which God communicates with mankind.[861]

The Chariot Vision and Scientific Method

In the interpretation given above that the Living Creatures are an analogy for psychological/moral decisions (See 'Middle Tier'), the Creatures may also be telling us something in terms of science and scientific decisions. If as discussed

---

[861] Galileo said that the book of the universe is written in the language of mathematics and its characters are triangles, circles and other geometric figures: '*La filosofia è scritta in questo grandissimo libro che continuamente ci sta aperto innanzi agli occhi (io dico l'universo), ma non si può intendere, se prima non s'impara a intender la lingua, e conoscer I caratteri ne' quali e scritto. Egli è scritto in lingua matematica, e I caratteri son triangoli, cerchi ed altre figure geometriche, senza I quali mezzi è impossibile intenderne umanamente parola.*'. Galilei, G. (1564).

above, the wheels are an analogy for science, and since 'the spirit of the living creatures was in the wheels' (Ez 1:20), then the balance between the Living Creatures may be an analogy for scientific method. It would seem therefore that four elements are crucial for science: the human face giving rationality (which is the most fundamental ingredient of scientific method), the eagle showing perspicacity or intuitive thinking (recognised as essential by Popper[862]), the lion showing authoritativeness (which could stress how important peer-review and confirmation of findings are in science) and finally the cattle giving the ingredient of nurturance (which might stress that ethical science must only be pursued for the benefit of mankind). The corollary of this is that the correct function of empirical science is also dependent on the spirit, not only for intuition and correct scientific method, but also in the interpretation and publication of the results.

Conclusions.

In summary, an attempt has been made in writing this thesis to put the text of Ezekiel into the context of what was an extraordinarily productive and precocious pre-scientific environment. Ezekiel found himself immersed in advances in mathematics and early calculus that were some 1,400 years ahead of their time. Given this context, it would not be surprising if Ezekiel's writing mirrored this fertile background of rationality-based pursuits.

Aided by the classical foundations of Maimondes' 'divine science' and Launderville's recent linking of reason with spirit, the conclusions reached here are justifiable by 'text in context' analysis: that reason and spirit work closely together in the Chariot Vision to illustrate the emotional revival of the Israelites. Reason and spirit also provide the co-ordinates of a pathway to right reason in a vision which also offers unexpected proto-scientific insights. While there are more speculative links with Christian theology which I treat in the coda to this

---

862   Popper, K. (1972), *The Logic of Scientific Discovery*, London: Hutchinson, p. 32.

work, most essentially, the fact that contemporary epistemology is reflected in the ancient three-tiered structure of Ezekiel 1 gives food for thought about the links between science and revelation. Although it may be contested that this is a fortuitous connection, I feel that the conclusions here are hermeneutically robust and demonstrate both the scientific as well as symbolic importance of this famous passage.

Conclusion of the Chapter

In the Bible, revelation that includes concrete objects may be illustrated using an intermediary more complex natural phenomenon. The Chariot Vision can be interpreted as linking Maimonides' 'divine science' and modern empirical science.

Chapter 13

Coda: The Trinitarian Signs – Science

Trinitarian Signs in the Chariot Vision

Following the interpretations of Block that see both Christ and the Holy Spirit in Ezekiel, there is a possible Trinitarian interpretation of the Chariot Vision. The appearance of the Glory of God of Israel as the 'likeness of a man', suggests Christ. The figure was seated on a throne which was on a 'firmament'. The appearance of the throne was like sapphire. Blue was the colour of Aaron's robe and is one of the three colours of blue, purple and scarlet of the priestly ephod. (Ex 39:1). Blue is therefore a holy colour in the Old Testament and the sapphire throne could be a representation of the presence of God the Father.[863] In the Old Testament, the 'God of Israel' appears on a pavement of sapphire stone (Ex 24:10), which has similarities to the appearance of the likeness of a man on the sapphire 'throne' in the Chariot of YHWH vision.

---

[863] The three colours of the Ephod, blue, scarlet and purple have a relationship to each other in that the combination of blue and scarlet gives purple. It is possible therefore that these colours could signify the three persons of the Holy Trinity with Scarlet representing Christ (the fire of the son of man figure), Blue representing God the Father, and purple, the Holy Spirit (as the Holy Spirit proceeds from Father and Son and this could be represented by a combination of the colours of the Father and the Son, blue and red, which gives purple. Purple *'argāmān* אַרְגָּמָן is a royal colour: Block, D. I. (1998), *The Book of Ezekiel Chapters 25–48*, Grand Rapids: Eerdmans, p. 61).

Block identifies the 'spirit' of the Chariot Vision with the Holy Spirit.[864] In the Chapter 'Spirit in Ezekiel and the Chariot' the centrality of the spirit to Ezekiel's message of renewal, and also the way the spirit links Ezekiel's visions was discussed. Block points out that when Christ tells Nicodemus that 'No one can enter the kingdom of God unless he is born of water and the spirit' (Jn 3:5), He is referring to Ez 36:26: 'a new spirit I will put within you'. In this passage from John therefore, Jesus himself identifies the Holy Spirit with 'the spirit' in Ezekiel.

Also, as noted above, the coals that burn between the Living Creatures, are the only concrete naturally occurring substance in the Chariot Vision. The coals could be an analogy of a sacrificial holocaust and the fire emanating and scattered over Jerusalem could be interpreted as the blood of the sacrifice. This sacrifice could be a foreshadowing of the sacrifice of the crucifixion and death of Christ and the pouring out of his blood, which would take place in Jerusalem.[865] This adds further symbolism to the Chariot Vision since YHWH says 'and you shall be my people and I shall be your God' (Ez 36:28) and the sacrifice of Golgotha is the way in which Christ draws everyone to himself, (Jn 12:32). If as noted above, the ḥašmal is a symbol of the 'heart of flesh' that YHWH will give the Israelites (Ez 36:26), then the coals and their scattering are a symbol of the process by which this 'new heart' will be given. Jacob of Serug was Bishop of Serug, Mesopotamia in the year AD 519. He wrote a homily on the Chariot Vision which has been preserved.[866] Jacob sees the Chariot as a portrayal of 'the holy altar of divinity'

The Son of God is all the beauty of prophecy,

---

[864] Block, D. I. (2013), *By the River Chebar: History, Literary, and Theological Studies in the Book of Ezekiel*, Eugene: Cascade, p. 15.

[865] cf. Is 31:9 'the LORD, whose fire is in Zion, and His furnace in Jerusalem.'

[866] Golitzin, A. (2003), 'The Image and Glory of God in Jacob of Serug's homily "On that Chariot that Ezekiel the Prophet saw",' *St Vladimir's Theological Quarterly*, 47, pp. 323–364; p. 355.

And without him there is neither prophecy nor revelation.

In the coals of fire are depicted the pearls of his body,

And in the chariot [is a portrait of] the holy altar of divinity

These coals of fire that are in the chariot depict his body.

They were placed in it both for retribution and for forgiveness.

One [of them] was given to Isaiah, to forgive his iniquity,

And [others] were scattered in Jerusalem for punishment.

Science and Religion

Science and the Bible

Although science developed in a religious atmosphere in the seventeenth century, modern science is largely secular. Science is usually seen to be at odds with religion and scientific findings are often construed as disproving the existence of God, or of showing that religion is redundant.[867] Showing that there are links between scientific method and the Scriptures is important to help refute this.

St Anselm's well-known statement: 'faith seeks understanding' and St Augustine's similar saying: *'Crede ut intelligas'* ('believe that you may understand') mean that to them, faith is required for and precedes understanding, but also that reason is essential to understanding. Rationality and logic therefore are not an independently reliable source of truth apart from the truth revealed in scripture.[868] Anselm however sees that even without scripture, the 'general revelation' of nature, should with reason, lead us to God.[869] This is also pro-

---

[867] Crosland, M. (2012), *Science under Suspicion*, Guildford: Grosvenor House, p. 61.

[868] Frame, J. M. (1983), 'Rationality and Scripture,' in Hart, H., van der Hoeven, J., Wolterstorff, N. (eds), *Rationality in the Calvinian Tradition*, Lanham (MD): University Press of America, pp. 293–317.

[869] Anselm, *Monologion*. 'If anyone does not know, either because he has not heard or because he does not believe, that there is one nature, supreme among all existing things,

claimed by Paul.[870] God can be immanent within the created order and can enter into intimate union with the world, because of his infinite difference from it.[871]

Scripture of itself is rational and precise.[872] Cherbonnier gives a compelling argument that that it was actually the Bible that allowed the development of modern scientific method. The Greek mind was convinced that the spatio-temporal world was not intelligible because it was mutable. The Greeks did not think true knowledge was attainable in the world of time and change; they thought that sense data was not reliable. They did not use inductive reasoning, but only deductive reasoning as they assumed all knowledge came from primary patterns and forms that were fundamental to how the world worked.[873] Greek science remained infertile until these basic premises of classical civilization had been overturned by those of the Bible: the Bible, firstly, showed that God is very vested in the temporal world. The Hebrews knew that the temporal world was an expression of God's action. This was completely against the premise that mutability was inferior to immutability and that change was a barrier to understanding things. Once it was understood that the temporal world is an expression of God's action, the world became intelligible in a way that it was not for the Greeks. Over time, mathematical tools including calculus, were developed by Newton, Leibniz and others and the door was opened

---

who alone is self-sufficient in his eternal happiness, who through his omnipotent good-ness grants and brings it about that all other things exist or have any sort of well-being, and a great many other things that we must believe about God or his creation, I think he could at least convince himself of most of these things by reason alone, if he is even moderately intelligent.'

[870] Ro 1:19 'for what can be known about God is plain to them because God has shown it to them.'

[871] Hanby, M. (2013), *No God, No Science? Theology, Cosmology, Biology*, Chichester: Wiley-Blackwell, p. 306.

[872] Thomas, S. J. (2004), 'The Rationality, Meaningfulness and Precision of Scripture,' *TMSJ*, 15 (2), pp. 175–207

[873] Cherbonnier, E. (1973), 'Biblical Contributions to the Development of the Scientific Method,' *The Modern Churchman*, 17 (1), p. 26.

for modern empirical science. Secondly, the Bible established the primacy of empirical observation. In the ancient world, perception was disparaged since abstract universals were thought to be the only worthwhile objects of study. The important biblical concept that changed this is that God is an individual, a concrete agent. This reality was coupled with the realisation that the only way to know an individual is through sense perception. This changed the human outlook about individual things. They were no more regarded as copies of ideals or patterns, but as entities with their own existence. This opened the door to studying naturally occurring objects, entities and phenomena. Thirdly, the world of the Bible is orientated toward inductive reasoning, not deductive. Kepler's insight that one had to work back to God's thoughts *after* him was a key to the necessity of inductive reasoning.[874] In the words of Collingwood:

> ... maintaining that God is omnipotent, and that the world of nature is a world of God's creation, completely altered this situation. It became a matter of faith that the world of nature should be regarded no longer as the realm of imprecision, but as the realm of precision. To say that a line in nature is not quite straight means for a Platonist that it is only an approximation to a straight line, the result of a praiseworthy but not altogether successful attempt on the part of some natural thing to construct a straight line or to travel in one...
>
> The line was drawn by God; and if God had wanted it to be straight, it would have been straight. To say that it is not exactly straight, therefore, means that it is exactly something else. The natural scientist must find out exactly what...[875]

The development of modern scientific method, depended on an outlook of the mind that not only saw change as inherent to the natural world, but saw God as the author of nature and diversity. Using inductive scientific method on

---

[874] Cherbonnier, E. (1973), *Ibid.* p. 29.

[875] Collingwood, R. G. (1940), *Essay on Metaphysics*, Oxford: Clarendon Press, p. 253.

diversity, enabled man to 'work back to God's thoughts after him'.[876] The Bible therefore being the revelation of God, incorporates the outlook of mind that favours the understanding of nature. In addition, what this current thesis aims to show is that not only are there close links between Science and Scripture, but a scientific structure is detectable in parts of Scripture. Collingwood's words are also an acknowledgement of the legitimacy of fuzzy sets and systems in understanding nature and the Bible. A corollary of the presence of 'science' in scripture is that when scientists search for new discoveries through empirical experimental structures, the technical objects that they discover are potentially new truths. Scientific findings that are found to be valid and confirmed by further research become discovered truths that have a reality independent of the scientific method.

Significance for Religion

Religion has for many years been on the defensive in terms of claims of its irrelevance to Science. 'The defining characteristic of modern philosophies is that...they attempt to exclude God'[877] This thesis should give ammunition against those claims by showing that some scriptural conclusions (such as: 'the purpose of the idealised temple is a message of hope for the restoration of Israel and YHWH as their God'[878]), are partly scientifically based. It should also establish the importance of science not only in its own right but also as an inextricable content of Scripture. That God can reveal through science should also be noted by those religions that repudiate rationality as an attribute of God.[879]

---

[876] Cf. Kepler, J. (1618), *Harmonices Mundi*.

[877] Hanby, M. (2013), *No God, No Science? Theology, Cosmology, Biology*, Chichester: Wiley-Blackwell, p. 299.

[878] Block, D. I. (1998), *The Book of Ezekiel Chapters 25-48*, Grand Rapids: Eerdmans, p. 746.

[879] Reilly, R. R. (2011), *The Closing of the Muslim Mind*, Wilmington: ISI Books, p.64.

Acknowledgements

I would like to thank my supervisors, Professor Anthony Towey and Dr Elisabetta Canetta for their time, interest, suggestions and insights without which I could not have written this thesis. I also thank Dr Tarcisius Mukuka who gave me important early guidance.

Bibliography

Aaboe, A. (1974), 'Scientific Astronomy in Antiquity in Hodson,' F. R. (ed.), *The Place of Astronomy in the Ancient World*, London: Oxford University Press.

Abelson, J. (2001), *Jewish Mysticism, An Introduction to the Kabbalah*, reprinted from 1913 version, Global Grey.

American Psychiatric Association (2013), *Diagnostic and Statistical Manual of Mental Disorders*, Fifth Edition (DSM-5), Arlington: American Psychiatric Association.

Anderson, J. R. (2007), *How Can the Human Mind Occur in the Physical Universe?*, New York: Oxford University Press.

Andrae, T. (1926), *Mystikens psykologi* (The Psychology of Mysticism), Stockholm: Diakonistyrelsen.

Baigent, M. (2005), *Astrology in Ancient Mesopotamia*, Rochester: Bear and company.

Biggs, C.R. (1996), *The Book of Ezekiel*, London: Epworth Press.

Biwul, J. (2013), *A Theological Examination of Symbolism in Ezekiel with Emphasis on the Shepherd Metaphor*, Carlisle: Langham.

Black, J., Green, A. (1992), *Gods, Demons and Symbols of Ancient Mesopotamia*, London: British Museum Press.

Blenkinsopp J. (1990), Ezekiel, *Interpretation: A Bible Commentary for Teaching and Preaching*, Louisville: John Knox Press.

Block, D. (1997), *The Book of Ezekiel chapters 1-24*, Grand Rapids: Eerdmans.

Block, D. (1998), *The Book of Ezekiel chapters 25–48* Grand Rapids: Eerdmans.

Block, D. I. (2013) *By the River Chebar: History, Literary, and Theological Studies in the Book of Ezekiel,* Eugene: Cascade.

Block, D. I. (2014), *Beyond the River Chebar: Studies in Kingship and Eschatology in the Book of Ezekiel,* Cambridge: James Clarke & Co.

Boadt, L. (1990), Ezekiel,*New Jerome Biblical Commentary,* London: Burns and Oates.

Boman, T. (1960), *Hebrew Thought Compared with Greek,* Trans. Jules L. Moreau, New York: Norton.

Bonaventure, St (1956), *The Journey of the Mind to God,* Indianapolis: Hackett.

Booth, W. (1983), *The Rhetoric of Fiction,* Chicago: University of Chicago Press.

Breasted, J. H. (2010), *The Edwin Smith Surgical Papyrus: Hieroglyphic Transliteration, Translation and Commentary,* **1**, Chicago: University of Chicago Press.

Brown, F., Driver, S., Briggs, C. (1991), *Brown-Driver-Briggs Hebrew and English Lexicon,* Oxford: Clarendon Press.

Brown, S. S., Fitzmeyer, J. A., Murphy, R. E. (1995), *The New Jerome Biblical Commentary,* London: Burns and Oates.

Brownlee, W. H. (1986,) Ezekiel 1–19, *World Biblical Commentary,* **28**, Waco: Word.

Carley, K.W. (1974), *The Book of the Prophet Ezekiel,* Cambridge: Cambridge University Press.

Carpenter, E., Thomson, D.L. (2010), *Ezekiel, Daniel,* Carol Springs: Tyndale House.

Cassirer. E. (1925), *Philosophie der symbolischen Formen II: Das mythische Denken,* Berlin: Bruno.

Cerny, J. (1957), *Ancient Egyptian religion.* London, Hutchinson's University Library.

Childs, B. S. (1979), *Introduction to the Old Testament as Scripture,* Philadelphia: Fortress Press.

Chisholm R. B. Jr. (1998), *From Exegesis to Exposition*, Grand Rapids: Baker.

Christman, A. R. (2005), *'What did Ezekiel See?,'* Leiden: Brill.

Clagett, M. (1999), *Ancient Egyptian Science. A Source Book*, Volume 3: Ancient Egyptian Science, Philadelphia: American Philosophical Society.

Clark, A. (2003), *Natural Born Cyborgs*, Oxford: Oxford University Press.

Clements, R. E. (1986): *The Chronology of the Redaction in Ezekiel 1–24*, pp 283–94 in Lust, J. (ed.), *Ezekiel and his Book: Textual and Literary criticism and their Interrelation*, Leuven: Leuven University Press.

Clements, R. E. (1996), *Ezekiel*, Westminster Bible Companion (WBC), Louisville: Westminster John Knox Press.

Cody, A. (1984), *Ezekiel*, Delaware: Michael Glazier.

Cohn-Sherbok, D., Cohn-Sherbok, L. (1994), *Jewish and Christian Mysticism*, New York, Continuum.

Cole, J. (ed.) (2014), *Didache Bible*, Ignatius Bible Edition, San Francisco: Ignatius Press,

Collé, E., Collé, C. (2013) (eds) *Sefer Yetzirah, the Book of Formation*, Short Version 1562, The Seven in One English-Hebrew Edition, Scotts Valley: Createspace.

Collingwood, R. G. (1940,) *Essay on Metaphysics*, Oxford: Clarendon Press.

Connaughton, L. (1980), *A–Z of the Catholic Church* Leigh-on-Sea: Kevin Mayhew.

Cooke, G. A. (1936), *The Book of Ezekiel*, Edinburgh: T. and T. Clark.

Cornill, C. H. (1886), *Das Buch des Propheten Ezechiel*, Liepzig: Hinrichs.

Crosland, M. (2012), *Science Under Suspicion*, 2nd edn, Guildford: Grosvenor House.

Damasio, A. R. (2005), *Descartes' Error: Emotion, Reason and the Human Brain*, New York: Harper Collins.

Dantzig, T. (2007), *Number: The Language of Science*, New York: Penguin Plume.

Davies, D. (2011), *Method and Metaphysics in Maimonides' Guide for the Perplexed*, Oxford: Oxford University Press.

Davis, E. F. (1989) *Swallowing the Scroll: Textuality and the Dynamics of Discourse in Ezekiel's Prophecy*, Bible and Literature Series 21, Sheffield: Almond.

Duguid, I. M. (1994), *Ezekiel and the Leaders of Israel*, Leiden: Brill.

Dürr, L. (1917), *Ezekiels Vision von der Erscheinung Gottes (Ez.c. 1 und 10) im Lichte der vorderasiatischen Altertumskunde*, Wurtzburg: Richter.

Fagothey, A.(1972), *Right and Reason*, 5th edn, St Louis: Mosby.

Fanthorpe, N. (2013), *Mysteries and Secrets of Numerology*, Toronto: Dundurn.

Fisch, S. (1950), *Ezekiel*, London: The Soncino Press.

Fodor, J. A. (1975), *The Language of Thought*, Cambridge, Mass.: MIT Press.

Fodor, J. A. (1983), *The Modularity of Mind*, Cambridge, Mass.: MIT Press.

Forsyth, D.R. (2020), *Making Moral Judgments: Psychological perspectives on morality, ethics and decision-making*, New York: Routledge.

Frame, J. M. (1983), 'Rationality and Scripture,' in Hart, H., Van Der Hoeven, J., Wolterstorff, N. (eds.), *Rationality in the Calvinian Tradition*, Lanham (MD): University Press of America, pp. 293–317.

Frankfort, H. (1946), 'Myth and Reality,' in Frakfort H. et al., *Before Philosophy*, Baltimore: Penguin Books.

Friberg, J. (2005), *Unexpected Links between Egyptian and Babylonian Mathematics*, London: World Scientific.

Galambush, J. (1991), *Jerusalem in the Book of Ezekiel: The City as Yahweh's Wife*, Altanta, Scholars' Press.

Galilei, G. (1864,) *Il Saggiatore*, Firenze: Barbèra.

Garfinkel, S. P. (1983), *Studies in Akkadian influences in the Book of Ezekiel*, Ann Arbor: University Microfilms.

Garscha, J. (1974), *Studien zum Ezechielbuch: Eine redaktionskritische Untersuchung von Ez 1–39*, Europäische Hoschsculchriften 23, Bern: Herbert Lang.

Genesius, H. W. F. (1857), *Hebrew–Chaldee Lexicon*, London: Bagster.

Gibson, J. J. (1986), *The Ecological Approach to Visual Perception*, Hillsdale: Lawrence Erlbaum.

Gilson, E. (1956), *The Christian Philosophy of St Thomas Aquinas*, Indiana: University of Notre Dame Press.

Glassner, J. J. (2003), *The Invention of Cuneiform*, Baltimore: Johns Hopkins University Press.

Goslin D. A. (ed.) (1969), *Handbook of Socialization Theory and Research*, Chicago, Rand Mc Nally.

Greenberg, M. (1983), *Ezekiel 1-20*, New York: Doubleday.

Hales, S. (1727), *Vegetable Staticks: or an Account of some Statical Experiments on the sap of Vegetables*, London: printed for W. & J. Innys and T. woodward.

Hanby, M. (2013), *No God, No Science? Theology, Cosmology, Biolog,*. Chichester: Wiley-Blackwell.

Hanson, N. R. (1958), *Patterns of Discovery*, Cambridge: Cambridge University Press.

Harris, A. (1979), *The World as I see it*, New York:, Philosophical Library.

Hausman, D. (1992), *The Inexact and Separate Science of Economics*, Cambridge: Cambridge University Press.

Hebra, A. (2014), *The Physics of Metrology*, New York: Springer.

Helmholtz, H. von (1930), *Counting and Measuring (1887)* Bryan, C.L. (trans.), New Jersey, Van Nostrand.

Hertz, H. (1999), *Die Constitution der Materie* (English: The Constitution of Matter), edited by Albrecht Fölsing), Berlin: Springer.

Hinke (1907), *A New Boundary Stone of Nebuchadrezzer I*, Philadephia: University of Pennyslvania.

Horowitz, W. (1998), *Mesopotamian Cosmic Geography*, Winona Lake: Eisenbrauns.

Howell, B. (2014), *In the Eyes of God: A Metaphorical Approach to Biblical Anthropomorphic Language*, Cambridge: James Clarke.

Høyrup, J. (2002), *Lengths, widths, surfaces: A portrait of old Babylonian algebra and its kin*, Sources and Studies in the history of mathematics and physical sciences, New York: Springer.

Hunger, H. (1992), *Astrological reports to Assyrian Kings*, State Archives of Assyria, Vol. VIII, Helsinki: Helsinki University Press.

Hunger, H., Pingree, D. E. (1999), *Astral Sciences in Mesopotamia*, Leiden: Brill

Jahn, G. (1905), *Das Buch Ezechiel auf Grund der Spetuaginta hergestellt, übersetzt und kritisch eklärt*, Leipzig: Pfeiffer.

Jellicoe, S. (1968), *The Septuagint and Modern Study*, Oxford: Clarendon Press.

Jerome, St (2017) *Commentary on Ezekiel, Ancient Christian Writers.* New York, The Newman Press.

Johnson, A. R. (1961), *The One and the Many in the Israelite Conception of God*, Cardiff, University of Wales Press.

Joyce, P. M. (2007), 'Ezekiel 40-42: The Earliest "Heavenly Ascent" Narrative?,' In de Jong H. J., Tromp, J. (eds), *The Book of Ezekiel and its Influence*, Aldershot: Ashgate.

Joyce, P. M. (2007), *Ezekiel; A Commentary*, New York: T and T Clark.

Kepler, J. (1618), *Harmonices Mundi*, Frankfurt: Gottfried Tampach.

Klee, R. (1997), *Introduction to the Philosophy of Science*, Oxford: Oxford University Press.

Knott, K. (2005), *The Location of religion: A Spatial Analysis*, London: Routledge.

Koch-Westenholz, U. (1995), *Mesopotamian Astrology: An Introduction to Babylonian and Assyrian Celestial Divination*, Copenhagen: Carsten Niebuhr Institute Near Eastern Studies.

Kreisel, H. (2015), *Judaism as Philosophy: Studies in Maimonides and the Medieval Jewish Philosophers of Provence*, Boston: Academic Studies Press.

Lakoff, G. Johnson, M. (1980), *Metaphors We Live By*, Chicago: Chicago University Press.

Lakoff, G., Johnson, M.(1999), *Philosophy in the Flesh*, New York: Perseus.

Lang, B. (1981), *Ezechiel: Der Prophet und das Buch (Erträge der Forschung)*, Darmstadt: Wissenschaftliche Buchgesellschaft.

Launderville, D. L. (2007,) *Spirit and Reason*, Waco: Baylor University Press.

Lévi-Strauss, C. (1962), *La Pensée Sauvage*, Paris: Plon.

Lucas, A. (1926), *Ancient Egyptian Materials*, New York: Longman.

Luckenbill, D. D. (1926), *Ancient Records of Assyria and Babylonia*, Chicago: University of Chicago Press.

Lull, J., Belmonte, J. A. (2009), 'Constellations of Ancient Egypt,' in Belmonte, J. A. Shaltout, M.,*In Search of Cosmic Order: Selected Essays on Egyptian Archaeoastronomy*, Cairo: Supreme Council of Antiquities Press.

Mackay, J.L. (2018), *Ezekiel. A Mentor Commentary. Volume 1: Chapters 1-24.* Fearn, Christian Focus Publications.

McKeating, H. (1993), *Ezekiel.* Sheffield, JSOT Press.

Maimonides, Moses (1956), *Mishneh Torah*, 6 Vols, New York: Rambam Publishers.

Matthews, I. G. (1939), *Ezekiel*, American Commentary on the Old Testament, 21, Chicago: Baptist Publication Society.

Mayr, E. (2004), *What Makes Biology Unique? Considerations on the autonomy of a scientific discipline*, Cambridge: Cambridge University Press.

Midrash Mishle (1884), f. 34a ff. ed. Cracow, Buber.

Migne, J.-P. (ed.) (1857–66), *Patrologiae Cursus Completus: Series Graeca*, Paris: Imprimerie Catholique.

Miller, J. H. (ed.) (1971), *Aspects of Narrative*, New York: Columbia University Press.

Moore, G. F. (1927), *Judaism*, 3 vols, Cambridge: Harvard University Press.

Odell, S. (2005), *Ezekiel*, Smyth & Helwys Bible Commentary, Macon: Smyth & Helwys.

Okasha, S. (2016), *Philosophy of Science: A Very Short Introduction*, Oxford: Oxford University Press.

Olyan, S. M. (2000), *Rites and rank : Hierarchy in Biblical Representations of Cult*, Princeton: Princeton University Press.

Origen (2010), *Homilies 1–14 on Ezekiel*, trans Sheck, T. P., New York: Newman Press.

Ortony, A. (1993) 'Metaphor, Language, and Thought,' in *Metaphor and Thought*, Ortony, A. ed., 2nd edn,. Cambridge: Cambridge University Press.

Patton, C. L., Cook, S. L. (eds) (2004), 'Ezekiel' in, *Ezekiel's Hierarchical World: Wrestling with a Tiered Reality*, Atlanta: Society of Biblical Literature.

Pedersen, O. (1993), *Early Physics and Astronomy: A Historical Introduction*, Cambridge: Cambridge University Press.

Peet, T. E. (1923), *The Rhind Mathematical Papyrus*, London: Hodder and Stoughton.

Petersen, B. N. (2012), *Ezekiel in Context*, Princeton Theological Monograph Series, Eugene: Wipf and Stock.

Petrie, F. (1883), *The Pyramids and Temples at Gizeh*, London: Field and Tuer.

Pinckaers, S. (1995), *The Sources of Christian Ethics*, Washington DC: The Catholic University of America Press.

Poincaré, H. (1903), 'Science and Method' in: Gould, S. J. (2001) *The Value of Science: Essential Writings of Henri Poincaré*, New York: Random House.

Popper, K. (1972), *The Logic of Scientific Discovery*, London: Hutchinson.

Porada, E. (1987), *On the Origins of Aquarius: Festschrift für Reiner*, Ann Arbor: American Oriental Society.

Poser R. (2012), *Das Ezechielsbuch als Trauma-Literatur*, Leiden: Brill.

Pritchard, J. B. (ed.) (1969), *Ancient Near Eastern Texts Relating to the Old Testament*, 3rd edn, Princeton: Princeton University Press.

Pritchard, J. B. (2011), *The Ancient Near East: An Anthology of Texts and Pictures*, Princeton: Princeton University Press.

Rad, G. von (2012,) *Old Testament Theology Volume Two: The Theology of Israel's Prophetic Traditions*, Louisville: Westminster John Knox Press.

Reddit, P. L. (2008), *Introduction to the Prophets*, Grand Rapids: Eerdmans,

Reilly, R. R. (2011) *The Closing of the Muslim Mind*, Wilmington: ISI Books.

Reiner, E., Pingree, D. E. (1981), *Babylonian Planetary Omens: Part 2* Malibu: Undena.

Reiner, E. Pingree, D.E. (1998) *Babylonian Planetary Omens. Part 3*. Groningen, Styx.

Rheinberger, H.-J. (1991), 'The "epistemic thing" and its technical conditions - from biochemistry to molecular biology,' in B. Gremmen (ed.), *The Interaction between Technology and Science*, Wageningen: Wageningen Agricultural University.

Robson, E. (2008), *Mathematics in Ancient Iraq: A Social History*, Oxford: Princeton University Press.

Rollins W. G,, Kille D. A. eds (2007), *Psychological Insight Into the Bible: Texts and Readings*, Grand Rapids: Eerdmans.

Ryan C. L. (2014), *Concerning Ministerial Visions: An Introduction to Ezekiel's Visions*, Vol. 2, Birmingham: Capital Cube Creatives.

Scholem, G. (1961), *Major Trends in Jewish Mysticism*, New York: Schocken.

Scholem, G. G. (1965), *Jewish Gnosticism, Merkavah Mysticism and Talmudic Tradition*, New York: The Jewish Theological Seminary of America.

Segal, R. E. (ed.) (1996), *Philosophy, Religious Studies and Myth*, New York: Routledge.

Seising, R. (2010), 'Fuzzy Sets and Systems and Philosophy of Science' in Seising R. (ed) *Views on Fuzzy Sets and Systems from Different Perspectives*, Hamburg: Springer.

Shepard, P. (1978), *Thinking Animals*, New York: Viking.

Siegman, E. (1961), *The Book of Ezechiel*, New York: Paulist Press.

Sirat, C. (1990), *A History of Jewish Philosophy in the Middle Ages*, Cambridge: University Press.

Sismondo, S. (2009), *An Introduction to Science and Technology Studies*, Chichester: Wiley.

Smend, R. (1880), *Der Prophet Ezechiel*, Liepzig: S.Hirzel.

Smith, J. Z. (1987), *To Take Place: Towards Theory in Ritual*, Chicago: University of Chicago Press.

Smith-Christopher, D. L. (2002), *A Biblical Theology of Exile*, Overtures to Biblical Theology, Minneapolis: Fortress Press.

Soggin, J. A. (1980), *Introduction to the Old Testament*, London: SCM Press.

Steinhart, E. (2001), *The Logic of Metaphor: Analogous Parts of Possible Worlds*, Berlin: Springer.

Strong, J. (2012), *Strong's Hebrew Dictionary of the Bible*, London: BN Publishing.

Sweeney, M. A. (2013), *Reading Ezekiel*, Macon: Smyth & Helwys.

Sweeney, M. A. (2014), *Reading Prophetic Books: Form, Intertextuality, and Reception in Prophetic and Post Biblical Literature*, Tübingen: Mohr Siebeck.

Taylor, J. B. (1969), *Ezekiel: An Introduction and a Commentary*, London: Tyndale.

Teresi, D. (2002), *Lost discoveries: The ancient roots of modern science – from the Babylonians to the Maya*, New York: Simon & Schuster.

Tillich, P. (2007), 'Religious Language as Symbolic' in *Philosophy of Religion: Selected Readings*, ed. Peterson, M. et al., Oxford: Oxford University Press

Torrey, C. C. (1970), *Pseudo-Ezekiel and the Original Prophecy*, New York: Ktav Publishing.

Wartofsky, M. W. (1968), *Conceptual Foundations of Scientific Though,*. New York: Collier-Macmillan.

Waerden, B. L. van der (1975), *Science Awakening*, Leiden: Noordhoff International.

Wevers, J. W. (1969), *Ezekiel*, London: Nelson.

White, G. (2007), *Babylonian Star-Lore*, London: Solaria.

Wilson, R. R. (1980), *Prophecy and Society in Ancient Israel*, Minneapolis: Fortress Press.

Wilson, R. R. (2000), 'Ezekiel", in *Harper Collins Bible Commentary*, Mays, J. L. *et al*. (eds), San Francisco: Harper San Francisco.

Wolff, H. W. (1974a), *A Commentary on the Prophet Hosea*, Philadelphia: Fortress Press.

Wolff, H. W. (1974b), *Anthropology of the Old Testament*, Philadelphia: Fortress Press.

Zadeh, L.A. (1987), 'Coping with the imprecision of the real world,' in R. R. Yager, S. Ovchinnikov, R. M. Tong, and H. T. Nguyen (eds), *Fuzzy Sets and Applications: Selected papers by Lotfi Zadeh*, New York: John Wiley.

Zimmerli, W. (1979), *Ezekiel 1*, Philadelphia: Fortress Press.

Zimmerli, W. (1983), *Ezekiel 2*, Philadelphia: Fortress Press.

Appendices

Appendix 1: Book of Enoch 14

[15] And as I quaked and trembled, I fell upon my face And I beheld a vision, And lo! There was a second house, greater [16] than the former, and the entire portal ftood open before me, and it was built of flames of fire. And in every respect it so excelled in splendour and magnificence and extent that I cannot describe it to [17] you its splendour and its extent. And its floor was of fire, and above it were lightnings and the path [18] of the ftars, and its ceiling also was also flaming fire. And I looked and saw therein a lofty throne: its appearance was as cryftal, and the wheels thereof as the shining sun, and there was the vision of [19] cherubim. And from underneath the throne came ftreams of flaming fire so that I could not look [20] thereon. And the Great Glory sat thereon, and His raiment shone more brightly than the sun and [21] was whiter than any snow. None of the angels could enter and could behold His face by reason [22] of the magnificence and glory, and no flesh could behold Him. The flaming fire was round about Him, and a great fire ftood before Him, and none around could draw nigh Him: ten thousand times [23] ten thousand (ftood) before Him, yet He needed no counsellor. And the moft holy ones who were [24] nigh to Him did not leave by night nor depart from him. And until then I had been proftrate on my face trembling: and the Lord called me with His own mouth and said to me: 'Come hither Enoch, and hear my word'

Appendix 2: Ezekiel Passages that mention the wheels of the Chariot of YHWH

Ez 1

[15] As I looked at the Living Creatures, I saw **a wheel** (*ofan*) on the earth beside the Living Creatures, one for each of the four of them. [16]As for the appearance of **the wheels** (*ofan*) and their construction: their appearance was like the gleaming of beryl; and the four had the same form, their construction being something like **a wheel within a wheel** *ofan*). [17] When they moved, they moved in any of the four directions without veering as they moved. [18] Their rims were tall and awesome, for the rims of all four were full of eyes all round. [19] When the Living Creatures moved, the **wheels** moved beside them; and when the Living Creatures rose from the earth, the wheels rose. [20] Wherever the spirit would go, they went, and the **wheels** rose along with them; for the spirit of the Living Creatures was in the **wheels**. [21] When they moved, the others moved; when they stopped, the others stopped; and when they rose from the earth, the **wheels** rose along with them; for the spirit of the Living Creatures was in the **wheels.**

Ez 3

[12] Then the spirit lifted me up, and as the glory of the Lord rose from its place, I heard behind me the sound of loud rumbling; [13] it was the sound of the wings of the Living Creatures brushing against one another, and the sound of the **wheels** beside them, that sounded like a loud rumbling.

Ez 10

[1 Then I looked, and above the dome that was over the heads of the cherubim there appeared above them something like a sapphire,* in form resembling a throne. [2] He said to the man clothed in linen, 'Go within the **wheel-work** (*galgal*) underneath the cherubim; fill your hands with burning coals from among the cherubim, and scatter them over the city.' He went in as I looked on. [3] Now the cherubim were standing on the south side of the house when the man went in; and a cloud filled the inner court. [4] Then the glory of the Lord rose up from the cherub to the threshold of the house; the house was filled with the cloud, and the court was full of the brightness of the glory of the Lord. [5] The sound of the wings of the cherubim was heard as far as the outer court, like the voice of God Almighty* when he speaks. [6] When he commanded the man clothed in linen, 'Take fire from within the **wheel-work**,(*galgal*) from among the cherubim', he went in and stood beside a **wheel**. [7] And a cherub stretched out his hand from among the cherubim to the fire that was among the cherubim, took some of it, and put it into the hands of the man clothed in linen, who took it and went out. [8] The cherubim appeared to have the form of a human hand under their wings. [9] I looked, and there were **four wheels** beside the cherubim, one beside each cherub; and the appearance of the **wheels** was like gleaming beryl. [10] And as for their appearance, the four looked alike, something like a **wheel within a wheel**. [11] When they moved, they moved in any of the four directions without veering as they moved; but in whatever direction the front **wheel** faced, the others followed without veering as they moved. [12] Their entire body, their rims, their spokes, their wings, and the **wheels—the wheels** of the four of them—were full of eyes all round. [13] As for the **wheels**, they were called in my hearing 'the **wheel-work**'(*galgal*) [14] Each one had four faces: the first face was that of the cherub, the second face was that of a human being, the third that of a lion, and the fourth that of an eagle. [15] The cherubim rose up. These were the Living Creatures that I saw by the river Chebar. [16] When the cherubim moved, the **wheels** moved beside them; and when the cherubim lifted up their wings to rise up from the earth, the **wheels** at their side did not veer. [17] When they stopped, the others stopped, and when they rose up, the others rose up with them; for the spirit of the Living Creatures was in them. [18] Then the glory of the Lord went out from the threshold of the house and ſtopped above the cherubim. [19] The cherubim lifted up their wings and rose up from the earth in my sight as they went out with the **wheels** beside them. They ſtopped at the entrance of the eaſt gate of the house of the Lord; and the glory of the God of Israel was above them. [20] These were the Living Creatures that I saw underneath the God of Israel by the river Chebar; and I knew that they were cherubim. [21] Each had four faces, each four wings, and underneath their wings something like human hands. [22] As for what their faces were like, they were the same faces whose appearance I had seen by the river Chebar. Each one moved ſtraight ahead.

Appendix 3: Key Characteriſtics of Objects as part of an Experimental Syſtem[880]

	BOUNDARY OBJECTS	EPISTEMIC OBJECTS	TECHNICAL OBJECTS
NATURE	*Concrete*	*Abstract*	*Concrete*
	One object that is differently interpreted and provides a holding ground for ideas for communication, translation and ſtandardization of meaning.	Characterized by lack and incompleteness; partially expressed in multiple inſtantiations; continuously evolving.	Ready-to-hand complete and unproblematic inſtruments.
OVER TIME	*Stable*	*In flux*	*Static*
	Stable enough to enable coordination across communities of practice.	Dynamic to enable knowledge work over time.	Fixed & ſtable tools
RELATIONS	*Multiple*	*Dyadic*	*Dyadic*
	Used in direct, cross-boundary interactions between multiple actors.	Particular inſtantiations are used by the expert subject.	Concrete inſtruments used by expert

---

880   Ewenſtein, B., Whyte, J. (2009), 'Knowledge Practices in Design: The Role of Visual Representations as "Epiſtemic Objects",' *Organiſation Studies*, **3**, (1), pp. 7–30, Sage Publications, p. 9.

'Fuzzy' Objects will be defined as those objects that are incomplete/uncertain including ordinary words or perceptions or phenomena that are not describable in precise scientific terms and that lend themselves to fuzzy analysis. They tend to follow criteria of biological phenomena.[881]

Appendix 4: Principal Theophanies in Scriptures (apart from Ezekiel)

*Ex 19:16**

So it came about on the third day, when it was morning, that there were **thunder** and **lightning flashes** and a **thick cloud** upon the mountain and a **very loud trumpet sound**.

Ex 19:18

Now Mount Sinai was all in smoke because the LORD descended upon it in **fire**; and its smoke ascended like the **smoke of a furnace**, and the whole mountain **quaked violently**.

Ex 20:18-20

All the people perceived the thunder and the lightning flashes and the sound of the trumpet and the mountain smoking;

*Re 11:19**

And the temple of God which is in heaven was opened; and the ark of His covenant appeared in His temple, and there were **flashes of lightning and sounds and peals of thunder and an earthquake and a great hailstorm**.

---

[881]   Mayr, E. (2004) What Makes Biology Unique? Considerations on the autonomy of a scientific discipline. Cambridge, Cambridge University Press.

Ex 3:2*

The angel of the LORD appeared to him in a **blazing fire** from the midst of a bush; and he looked, and behold, the bush was burning with fire, yet the bush was not consumed.

Ex 13:21*

The LORD was going before them in a pillar of **cloud** by day to lead them on the way, and in a pillar of **fire** by night

Ex 24:17*

And to the eyes of the sons of Israel the appearance of the glory of the LORD was like a consuming **fire** on the mountain top.

Le 9:23–24*

Then fire came out from before the LORD and consumed the burnt offering and the portions of fat on the altar;

Jd 13:20–22*

For it came about when the flame went up from the altar toward heaven, that the angel of the LORD ascended in the flame of the altar.

Is 6:4*

And the foundations of the thresholds trembled at the voice of him who called out, while the temple was filling with smoke.

Re 15:8*

And the temple was filled with smoke from the glory of God and from His power;

Ex 24:16

The glory of the LORD reſted on Mount Sinai, and the cloud covered it for six days; and on the seventh day He called to Moses from the midſt of the cloud.

Ex 34:5

The LORD descended in the cloud and ſtood there with him as he called upon the name of the LORD.

De 31:15*

The LORD appeared in the tent in a pillar of cloud, and the pillar of cloud ſtood at the doorway of the tent.

Nu 9:15–22*

Now on the day that the tabernacle was erected the cloud covered the tabernacle, the tent of the teſtimony, and in the evening it was like the appearance of fire over the tabernacle, until morning.

1 Ki 8:10–11*

It happened that when the prieſts came from the holy place, the cloud filled the house of the LORD,

1 Ki 8:11

so that the prieſts could not ſtand to miniſter because of the cloud, for the glory of the LORD filled the house of the LORD.

*Mt 17:5**

While he was still speaking, a bright cloud overshadowed them, and behold, a voice out of the cloud said, "This is My beloved Son, with whom I am well-pleased; listen to Him!"

Mk 9:7

Then a cloud formed, overshadowing them, and a voice came out of the cloud, "This is My beloved Son, listen to Him!"

Lk 9:34–35

While he was saying this, a cloud formed and began to overshadow them; and they were afraid as they entered the cloud. Then a voice came out of the cloud, saying, "This is My Son, My Chosen One; listen to Him!"

*Re 14:14–16**

Then I looked, and behold, a white cloud, and sitting on the cloud was one like a son of man, having a golden crown on His head and a sharp sickle in His hand.

Is 6:1

In the year of King Uzziah's death I saw the Lord sitting on a throne, lofty and exalted, with the train of His robe filling the temple.

*Da 7:9**

"I kept looking Until thrones were set up, And the Ancient of Days took His seat; His vesture was like white snow And the hair of His head like pure wool His throne was ablaze with flames, Its wheels were a burning fire.

Is 6:2

Seraphim stood above Him, each having six wings: with two he covered his face, and with two he covered his feet, and with two he flew.

Re 4:6–11

and before the throne there was something like a sea of glass, like crystal; and in the centre and around the throne, four Living Creatures full of eyes in front and behind. The first creature was like a lion, and the second creature like a calf, and the third creature had a face like that of a man, and the fourth creature was like a flying eagle. And the four Living Creatures, each one of them having six wings, are full of eyes around and within; and day and night they do not cease to say, "HOLY, HOLY, HOLY IS THE LORD GOD, THE ALMIGHTY, WHO WAS AND WHO IS AND WHO IS TO COME."

Ex 24:10

and they saw the God of Israel; and under His feet there appeared to be a pavement of sapphire, as clear as the sky itself.

Re 4:3*

And He who was sitting was like a jasper stone and a sardius in appearance; and there was a rainbow around the throne, like an emerald in appearance.

Ex 40:34–35*

Then the cloud covered the tent of meeting, and the glory of the LORD filled the tabernacle. Moses was not able to enter the tent of meeting because the cloud had settled on it, and the glory of the LORD filled the tabernacle.

2 Ch 7:1–3*

Now when Solomon had finished praying, fire came down from heaven and consumed the burnt offering and the sacrifices, and the glory of the LORD filled the house.

Nu 12:9–10

So the anger of the LORD burned against them and He departed. But when the cloud had withdrawn from over the tent, behold, Miriam was leprous, as white as snow. As Aaron turned toward Miriam, behold, she was leprous.

Nu 12:5–8

Then the LORD came down in a pillar of cloud and stood at the doorway of the tent, and He called Aaron and Miriam.

Appendix 5: Ez 47: 1–6

Then he brought me back to the entrance of the temple; there, water was flowing from below the threshold of the temple towards the east (for the temple faced east); and the water was flowing down from below the south end of the threshold of the temple, south of the altar. [2] Then he brought me out by way of the north gate, and led me round on the outside to the outer gate that faces towards the east;* and the water was coming out on the south side. [3] Going on eastwards with a cord in his hand, the man measured one thousand cubits, and then led me through the water; and it was ankle-deep. [4] Again he measured one thousand, and led me through the water; and it was knee-deep. Again he measured one thousand, and led me through the water; and it was up to the waist. [5] Again he measured one thousand, and it was a river that I could not cross, for the water had risen; it was deep enough to swim in, a river that could not be crossed. [6] He said to me, 'Mortal, have you seen this?'

Appendix 6: Revelation 6: 5–6

When he opened the third seal, I heard the third living creature call out, 'Come!.' I looked, and there was a black horse! Its rider held a pair of scales in his hand, and I heard what seemed to be a voice in the midst of the four Living Creatures saying, 'A quart of wheat for a day's pay, and three quarts of barley for a day's pay, but do not damage the olive oil and the wine!'

Appendix 7: Three level classification of Scripture Passages

(a) Three level Classification of objects in Ezekiel Water flowing from the Temple Vision

(objects with an asterisk* have been selected for the fuzzy epistemological analysis)

UPPER LEVEL	MIDDLE LEVEL	LOWER LEVEL
Voice commanding Ezekiel*	The river*	The measuring line*
(God/an angel)	Many trees on river bank	The Temple thresholds
		The outer gate

(b) Three level Classification of the Black Horse vision in Re 6.5.

(objects with an asterisk* have been selected for the fuzzy epistemological analysis)

UPPER LEVEL	MIDDLE LEVEL	LOWER LEVEL	NOT ASSIGNABLE
Voice saying 'Come'*	flashes of lightning, thunder (Re 4.5)	pair of scales*	black horse
		seal	rider
	'A quart of wheat... olive oil and wine'*	scroll (Re 5.1)	lamb (Re 6.1), living creature

Appendix 8: Three-Level Classification of Objects in Ezekiel Chariot Vision of YHWH (Ez 1.4)

(objects with an asterisk* have been selected for the fuzzy epistemological analysis)

UPPER LEVEL	MIDDLE LEVEL	LOWER LEVEL	NOT ASSIGNABLE
Gleaming amber (Hašmal)	Stormy wind	wheels*	four Living Creatures
The spirit*	great cloud	coals	firmament of crystal
Sound of thunder of	brightness		likeness of a throne
Almighty Voice above	flashing fire		likeness of a
the firmament	burning coals		human form
	lightning*, rainbow		

Appendix 9: Movements of the Chariot of YHWH

CHAPTER	MOVEMENT
1:4	(Descent of Chariot Vision from the cloud to river bank)
1:4	Darting back and forth. Went, stood and rose
3:12	No movement mentioned
3:23	Stood
8:2	No movement mentioned
8:4	No movement mentioned
9:3	Went up from cherubim to the threshold of the Temple.
10:3	Standing on the south side of the Temple.
10:4	Went up from the cherubim to the threshold of the Temple.
10:18	'Glory of the Lord went forth from the threshold of the Temple. Mounted up from the earth and stood at the door of the East Gate'
11:23	Went up from the midst of the city and stood upon the mountain which is on the East side of City.
43:2	The glory of the God of Israel came from the east.
43:4	The glory of the Lord entered the Temple by the gate facing east.

Appendix 10: Post Traumatic Stress Disorder[882]

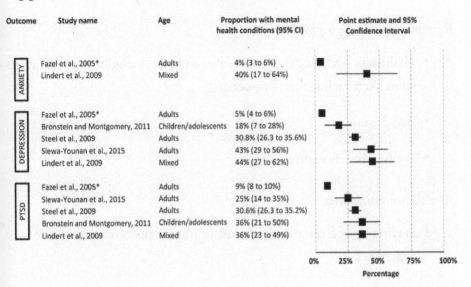

Outcome	Study name	Age	Proportion with mental health conditions (95% CI)	Point estimate and 95% Confidence Interval
ANXIETY	Fazel et al., 2005*	Adults	4% (3 to 6%)	
	Lindert et al., 2009	Mixed	40% (17 to 64%)	
DEPRESSION	Fazel et al., 2005*	Adults	5% (4 to 6%)	
	Bronstein and Montgomery, 2011	Children/adolescents	18% (7 to 28%)	
	Steel et al., 2009	Adults	30.8% (26.3 to 35.6%)	
	Slewa-Younan et al., 2015	Adults	43% (29 to 56%)	
	Lindert et al., 2009	Mixed	44% (27 to 62%)	
PTSD	Fazel et al., 2005*	Adults	9% (8 to 10%)	
	Slewa-Younan et al., 2015	Adults	25% (14 to 35%)	
	Steel et al., 2009	Adults	30.6% (26.3 to 35.2%)	
	Bronstein and Montgomery, 2011	Children/adolescents	36% (21 to 50%)	
	Lindert et al., 2009	Mixed	36% (23 to 49%)	

Figure 25: Prevalence rates (%, with 95% CI) as reported by systematic reviews calculating overall summary measures. *A 99% confidence interval was used by Fazel, M. *et al.* (2005), 'Prevalence of serious mental disorder in 7000 refugees resettled in western countries: a systematic review,' *Lancet*, **365** (9467), pp. 1309–1314.

Diagnosis of Post-Traumatic Stress Disorder [883]

To be diagnosed with PTSD, an adult must have all of the following for at least 1 month:

- At least one re-experiencing symptom

- At least one avoidance symptom

- At least two arousal and reactivity symptoms

---

[882] Turrini, G. et al. (2017) Common mental disorders in asylum seekers and refugees: umbrella review of prevalence and intervention studies. *Int J Ment Health Syst* 11:51. Figure 2.

[883] 'Post Traumatic Stress Disorder' at *https://www.nimh.nih.gov/health/publications/post-traumatic-stress-disorder-ptsd/index.shtml* accessed 23/10/2018.

- At least two cognition and mood symptoms

Re-experiencing Symptoms:

- Flashbacks – reliving the trauma over and over, including physical symptoms like a racing heart or sweating

- Bad dreams

- Frightening thoughts

Avoidance symptoms:

- Staying away from places, events, or objects that are reminders of the experience

- Avoiding thoughts or feelings related to the traumatic event

Arousal and reactivity symptoms:

- Being easily startled

- Feeling tense or 'on edge'

- Having difficulty sleeping, and/or **having angry outbursts**

Arousal symptoms are usually constant, instead of being triggered by something that brings back memories of the traumatic event. They can make the person feel stressed and angry.

Cognition and Mood Symptoms:

- Trouble remembering key features of the traumatic event

- **Negative thoughts about oneself or the world**

- Distorted feelings like guilt or blame

- Loss of interest in enjoyable activities

Cognition and mood symptoms can begin or worsen after the traumatic event. These symptoms can make the person feel alienated or detached from friends or family members.

Appendix 11: Bible Weights and Measures[884]

Lengths

NAME	AVOIRDUPOIS	METRIC	REFERENCE
Finger	0.73 in	1.85 cm	Je 52:21
Handbreadth (4 finger)	2.92 in	7.4 cm	Ex 25:25
Span	9 in	22.86 cm	Ex 28:16
Cubit	18 in	45.72 cm	Mt 6:27
Long Cubit	21 in	53.34	Ez 40:5
Fathom	6 feet	1.829 mi	Ac 27:28
Reed(6 long cubits)	10.5 ft	3.20	Ez 40:5
Furlong (KJV)	1/8 mi, 660 feet	201.168 m	Re 14:20
Stadion	607 ft	185 m	Lk 24:13
Sabbath day's journey (2,000 cubits)	3,000 ft	914 m	Ac 1:12
Day's journey	20 miles	32.187 km	1Ki 19:4
Gerah (1/20 shekel)	0.0201 oz	0.57 g	Ez 45:12
Beka (1/2 shekel)	0.201 oz	5.70 g	Ge 24:22
Pim (2/3 shekel)	0.268 oz	7.60 g	1 Sa 13:21
Shekel (20 gerahs)	0.402 oz	11.4 g	Ex 30:23
Mina (50 shekels)	1.256 lb	0.57 kg	Ezra 2:69
Talent (60 minas)	75.4 lb	34.2 kg	Ezra 8:26

---

[884]   From *https://biblehub.com/weights-and-measures/* accessed 8/3/2020.

Weights

NAME	AVOIRDUPOIS	METRIC	REFERENCE
Gerah (1/20 shekel)	0.0201 oz	0.57 g	Ez 45:12
Beka (1/2 shekel)	0.201 oz	5.70 g	Ge 24:22
Pim (2/3 shekel)	0.268 oz	7.60 gr	1Sa 13.21
Shekel (20 gerahs)	0.402 oz	11.4 g	Ex 30:23
Mina (50 shekels)	1.256 lb	0.57 kg	Ezra 2:69
Talent (60 minas)	75.4 lb	34.2 kg	Ezra 8:26

Liquid Measures

NAME	AVOIRDUPOIS	METRIC	REFERENCE
Log	0.33 quarts	0.31 l	Le 14:10
Cab (4 logs)	1.32 quarts	1.25 l	2 Ki 6:25
Hin (12 logs)	0.98 gal	3.7 l	Nu 15:4
Bath (6 hins)	5.8 gal	22 l	Is 5:10
Homer (10 baths)	58 gal	220 l	Ez 45:11
Cor (10 baths)	58 gal	220 l	Ez 45:11
Metretes (NT)	10.4 gal	39.4 l	Jn 2:6
Bath (NT)	8.7 gal	33 l	Lk 16:6

Dry Measures

NAME	AVOIRDUPOIS	METRIC	REFERENCE
Cab (1/18 ephah)	1.1 dry quarts	1.2 l	2 Ki 6:25
Omer (1/10 ephah)	2.0 dry quarts	2.2 l	Ex 16:36
Seah (1/3 ephah)	6.7 dry quarts	7.3 l	2 Ki 7:1
Ephah (10 omers)	0.624 bushels (5 dry gal)	22 l	Ru 2:17
Lethech (5 ephaths)	3.12 bushels (25 dry gal)	110 l	Ho 3:2
Homer (10 ephaths)	6.24 bushels (50 dry gal)	220 l	Le 27:16
Cor (10 ephaths)	6.24 bushels (50 dry gal)	220 l	Ez 45:14
Cor (NT)	10 bushels (93 dry gal)	350 l	Lk 16:7

Money

COIN	VALUE	VALUE METRIC	REFERENCE
Denarius	day's wage		Mt 20:2
Drachma	est. 0.035 oz. silver	est. 1 g silver	Lk 15:8
Didrachma (2 drachma)	est. 0.07 oz. silve	est. 2 g silver	Mt 17:24
Talent, silver	~ 75.4lb	~ 34.24 kg	Ezra 8:26
Talent, gold	~ 75.4lb	~ 34.24 kg	1 Ki 9:28

Appendix 12: Typical Traits of the 'Living Creatures'

	BABYLONIA	SCRIPTURE	MODERN
HUMAN:	wisdom (king) fertility, bringer of water and rain*[a]	most noble creature, image of God Son of God	rationality, wisdom, ability to distinguish good and evil
EAGLE:	takes dead to stars[b] swiftness all-seeing (king) lofty, proud, free	swiftest and most stately[c] lofty, spiritual[d]	far-sighted freedom lofty
LION:	king of beasts (king) killer of man and livestock. protective figure/ beneficial demon power of the sun	ferocity, courage[e] royalty (Judah, David) Ferocious.	fearlessness leadership imperial
BULL:	strength*[f] fertility (king) prosperity[g]	slaughter, sacrificial/priestly reproduction.	strong-willed, powerful fertile
COW:	celestial[h] fertility,	prosperity birth,	soft, nurturing, motherhood[i]
OX:	wealth	valuable domesticated animal[j] fertility and divinity symbol[k]	

---

[a] Constellation 'Gula': The Great One.

[b] White, G. (2007), *Babylonian Star-Lore,* London: Solaria, p. 126.

[c] De 28:49; Is 40:31; Je 48:40.

[d] Christman, A.R. (2005) *'What did Ezekiel See?,'* Leiden: Brill, p. 17.

[e] Ju 14:18; Sa 1:23, Sa 17:10.

[f] The Aurochs: extinct bull.

[g] White, G. (2007), *Babylonian Star-Lore,* London: Solaria, p. 84.

[h] White, G. (2007), *ibid.,* p. 84.

[j] Pr 14:4

[k] Ps 106:19.

TIGER OF THE STRIPE

TYPESET IN THE UK BY
TIGER OF THE STRIPE
USING LUALATEX
WITH EB GARAMOND
FOR THE MAIN TEXT,
EZRA SIL SR
FOR HEBREW
& SCHEHERAZADE
FOR ARABIC

Ingram Content Group UK Ltd.
Milton Keynes UK
UKHW041338140623
423364UK00002B/23

9 781904 799740